D0371568

THE
LAUGHING
CURE

THE
LAUGHING
CURE

Emotional and Physical Healing—A Comedian
Reveals Why Laughter Really Is the
Best Medicine

DR. BRIAN KING

Skyhorse Publishing

Skyhorse Publishing books may be purchased in bulk at special discounts for sales promotion, corporate gifts, fund-raising, or educational purposes. Special editions can also be created to specifications. For details, contact the Special Sales Department, Skyhorse Publishing, 307 West 36th Street, 11th Floor, New York, NY 10018 or info@skyhorsepublishing.com.

Skyhorse® and Skyhorse Publishing® are registered trademarks of Skyhorse Publishing, Inc.®, a Delaware corporation.

Visit our website at www.skyhorsepublishing.com.

10 9 8 7 6 5 4 3 2 1

Library of Congress Cataloging-in-Publication Data is available on file.

Cover design by Rain Saukas

Print ISBN: 978-1-5107-0249-3
Ebook ISBN: 978-1-5107-0250-9

Printed in the United States of America

For my mother, who taught me how to laugh, and Sarah, who makes sure that I laugh everyday.

CONTENTS

Introduction 1
What Are We Talking About, Anyway? 15
So What Makes Some Things Funny? 21
Humor: So What's the Point? 28
What Is a Joke? 35
Jokes: Again, What's the Point? 44
The Brain 46
The Brain: What's the Point? 53
Speaking of Laughter . . . 54
Laughter: What's the Point? 66
The Benefits of Laughter: Why We Should Care 69
Laughter Makes Us Happy 72
Okay, So What's So Important about Happiness Anyway? 83
Laughter Feels Good 88
Do We Really Need to Discuss the Importance
  of Pleasure? 91
Laughter Decreases Pain 92
Is No Pain Our Gain? 102
Laughter Reduces Stress 107
Other Than the Obvious, Why Is Stress Management
  So Important? 122
Laughter Is Healthy 124
Comedians with Diabetes on Coping with Diabetes 133

Okay, So What Is So Important about, Um … Health?   140

Laughter Is Exercise   144

Laughter Makes Us Friends (and More)   146

Now That the Question Has Been (Sort of) Answered …   160

What Is So Important about Making Friends?   162

Laughter Is Foreplay   166

Why Is Foreplay Important?   171

Laughter Helps Sustain Relationships: My Chat
   with Yakov   172

Laughter Helps Calm Our Mind   178

Why Is It So Important to Overcome Negative Thinking?   189

Laughter Generates More Power Than Screams   192

The Benefits of Laughter: A Recap   193

Okay, So What Can We Do with This Information?   195

Learn How to Not Take Life Too Seriously   200

Learn How To tell a Damn Joke   206

Recognize the Therapeutic Value of Humor and Laughter   214

Practice Laughter, and Laugh Often   222

Final Thoughts   227

Appendix I: Humor Testimonials   230

Appendix II: Expressing Gratitude   246

Endnotes   248

Acknowledgments   259

INTRODUCTION

Hi.

I've always thought that a simple "hi" is the best way to break the ice with someone, and as I am introducing myself to you it seemed like a natural, fully appropriate way to kick off this paragraph. After all, I am a very simple guy; I have simple needs and values, laugh at stupid things, and make a living out of simplifying ridiculously complex concepts to make them more comprehensible to average, simple guys like myself. People needlessly complicate their lives and too often writers use elaborate, colorful language or take entirely too long to get to their point (see what I did there?). So, hi.

My name is Dr. Brian King, and I am both a psychologist and a stand-up comedian. A lot of people think that's an unusual combination; but actually, I know of several people, including a few personally, who could claim the same thing. My specific brand of psychology is called biopsychology, a term synonymous with behavioral neuroscience, which you're more likely to have heard of. It's the study of the brain as the basis for behavior. Biopsychology and comedy is a combination that is more unusual than general psychology and comedy, and I used to think I was the only combiner of them. This changed a few years ago when I performed in a show in Austin, Texas. Austin is one of the greatest cities our society has ever produced and home to my college alma mater, the University of Texas. The comedian before me, Juan Salinas, was a professor of neuroscience at the school, and thanks to him, I can no longer call myself the only stand-up comedian trained in neuroscience. Coincidentally, it turned out that Dr. Salinas even received his PhD with two future professors who later helped me get mine at the University of New Orleans. It seems that comedy and neuroscience are small worlds. However, I don't normally identify as a neuroscientist; I think of myself more as a general psychologist and that's what I'll refer to myself as going forward.

I like to think that comedy and psychology are highly related to one another. For example, as a psychologist I make observations on human behavior. As a comedian, I make fun of it. Psychologists like to point out that everyone, by virtue of being a member of this species, is an amateur psychologist. This is especially true of comedians, who observe behavioral phenomena that the rest of us might miss and come up with commentary or an explanation, a hopefully funny one. Comedians are master communicators and are very sensitive to body language, nonverbal cues, and other subtleties in the same way that a good therapist is attuned to clients. Writing good comedy requires insight into the human experience, not just for the mining of material, but also to be relatable to an audience. I think the two fields definitely draw from the same fundamental skill set.

Depending on your point of view, I'm either a psychologist who performs comedy or a comedian who happens to have a doctorate. I think of myself as the latter, and believe it or not, getting my PhD was my *backup* plan. As far back as I can remember, I always wanted to be a comedian. I started thinking about it while in college in Austin, so I guess that's as far back as I can remember. At the risk of dating myself, I was there when a comedy club called the Velveeta Room opened up on East 6th Street and I would pop in on a regular basis. I became familiar with some of the local comedians and they urged me to give it a shot, and holy crap I wanted to, but something held me back. Then I moved, and moved again, and again, and a few more times in pursuit of my PhD and later my struggling psychology career. Years later, I was living in San Francisco and walked past a sign for the "San Francisco Comedy College"[1] and a light bulb went off. I had no idea such a place existed, and although I don't believe in fate, I felt that I was definitely in the right place at the right time of my life. I signed up for the very next class I could, and was hooked on performing from that moment on.

Comedy felt right. Looking back on my life, the only thing I have ever consistently been able to do well is make people laugh. I was a class clown and entertained my family. I made friends through laughter, filmed comedy skits in graduate school to relieve stress, and was a decent professor mainly because my students enjoyed my sense of humor. After I had been doing stand-up for a few years, I got a call from someone who organizes continuing education courses. He had heard about my experience in comedy and my background in psychology and thought marketing seminars taught by a psychologist/comedian might generate additional interest. I agreed and started touring as a comedic public speaker and it's proven to be one of the best decisions I've ever made. I now give seminars on happiness, humor, and health to thousands of people all over the country each year. Combining psychology and comedy even led to my having the opportunity to write this book.

Oh, and despite my introduction and the cover, I don't regularly go by "Dr. Brian King"—that's my stage name. It helps me distinguish myself from all the other white, fat, male comedians in the world (there are a lot of us out there). I can't say it's made me famous, but I have been identified on the street as "that doctor comedian" more than once, and no, it wasn't by my mom. You can call me Brian.

Now how was that for a simple introduction?

As a psychologist, I am really interested in ways we can make life better. As a comedian, I am really interested in making people freaking laugh, and laughter makes life better. Laughter relieves stress. It does, trust me on this for now, but it does. In this book I'm going to discuss the importance of having a sense of humor and the impact it has on our mental and physical health. Enjoying and appreciating a good laugh, recognizing humor in life, and most importantly not taking life too seriously is enormously beneficial. Life is a serious subject, seriously it is, but paradoxically the very serious lesson is that we should not take it *too* seriously, so please don't. I don't, and I'm a professional.

Let's begin with the psychological benefits of humor. How does humor help us emotionally?

I've often thought of this as a good question while smugly complimenting my own inquisitiveness. Let us frame the discussion of taking life less seriously by first discussing the opposite. There are a handful of psychological disorders, most notably clinical depression and anxiety, that we collectively refer to as emotional disorders. These are HORRIBLE disorders. They severely impact the quality of life of the individuals who suffer from them and also that of the people around them—their family, friends, peers, classmates, and coworkers—and also the greater society at large. Depression currently places the largest burden on our society in terms of reduced years of healthy life,[2] and we lose a tremendous amount of productivity to depression, compensating for it and recovering from it.[3] It impacts our immune system[4] and is a factor in the experience of chronic pain.[5] Finally, it has disastrous ultimate consequences if an individual unfortunately succumbs to suicide as a result of the condition. Emotional disorders are an extremely serious topic, and one that I do not often joke about.

In my seminars, I am very fond of pointing out the importance of humor to overall happiness. I want you to consider something for a moment, think about how you respond to humor: what are the behavioral expressions of humor? When most people find something funny, they'll smile or laugh. Duh.

Now, think about the behavioral expressions of being happy: when you are happy, how do you express it? Well, most people express happiness through smiling and laughter. It's the same physiological response. Smiles and laughter are both a reaction to appreciating humor and a response to feeling happy.

And it's not just humor and happiness that provoke a similar response, smiling and laughing do too; they both make us happier. William James, the father of American psychology (and probably the originator of the crazy, out-of-control beard look so stereotypical of psychologists), once said, "We don't laugh because we're happy, we're happy because we laugh." What he was referring to has since become one of the most robust theories in psychology and does a really good job of describing the experience of all emotion. Basically, our emotions are experienced as the brain's interpretation of physiological changes in our body. Let's say we encounter some stimulus, like a snarling wolf, or our favorite cousin (hey, Doug!). When we encounter any stimulus, the first thing our brain needs to do is make sure it isn't going to kill us. Areas of our brain that lie outside our awareness will process the stimulus and form an immediate initial reaction. That leads to a physiological response that might include things like our heart rate increasing, muscular tension, or the activation of our tear ducts. That reaction is automatic; it occurs prior to our conscious awareness and gets us ready for an appropriate behavioral action. That physiological reaction provides feedback to the prefrontal cortex, the part of your brain that integrates information, and combined with the information about the stimulus, is used to determine the emotional experience. Our prefrontal cortex might receive information that our heart rate has increased and we are in the presence of a threatening wolf and interpret that combination of inputs as the emotion of fear, or joy in the case of seeing Doug (until we later remember that he owes us money). In other words, our body informs our brain as to what emotion it is experiencing. Our brain receives feedback from the body regarding its emotional condition. Again, we aren't laughing because we are happy, we're happy because we're laughing. Over time, this has become known as the "James–Lange theory" of emotion because William wasn't the only one to describe this phenomenon—he shares the credit with Carl Lange.[6] Smiling and laughter make us happy or, to state it more scientifically, increase our experience of positive affect. It's not just theory; there are lots of studies that show we can improve affect through engaging in behaviors our brain associates with feeling happy. For example, Fritz Strack, Leonard Martin, and Sabine Stepper (1988)[7] showed that research participants tricked into smiling involuntarily not only felt happier as a result but also found cartoons more amusing. They had participants hold a pen in their mouth in one of two ways: either with their lips holding it so that it stuck straight out like a cigarette, which created a facial configuration similar to a frown, or sideways in their teeth, which made their mouth look a lot like a smile. Their simple manipulation has become my favorite demonstration of the James–Lange theory because who doesn't have a pen on hand? It's also my favorite intervention for negative affect; if you find yourself feeling down one day and there's nothing you can do to feel better, PUT A PEN IN YOUR MOUTH. It can't hurt. Just don't bite down too hard.

Just as acting happy makes us happy, acting miserable makes us miserable. Our physiology can contribute to negative emotion as well. Generally, when we conduct research on human participants, we don't want to cause any permanent harm as a result of our studies (we kind of depend on them coming back), so we won't do anything to make them more miserable than they already might be. In other words, we won't ask them to do something that we hypothesize may cause a negative response. However, we can learn a lot by studying the effects of eliminating negative responses. Maybe it's just because I live in Hollywood, but I think most people are familiar with the botulinum toxin, commonly called

Botox. When injected into the face, it paralyzes facial muscles, and people pay good money for this service because apparently paralyzed facial muscles are very attractive (judging by its popularity in my neighborhood, I may have chosen to be the wrong type of doctor). Well, Michael Lewis and Patrick Bowler (2009)[8] gave frown-inhibiting injections of Botox to participants in their study, and those participants reported feeling happier and less anxious. This effect wasn't due to the cosmetic benefit of the injections; participants did not report feeling more attractive, but most likely felt better because *they could no longer frown.*

Humor helps relieve anxiety and depression, but it has benefits beyond our emotions. Remember earlier when I said that humor helps relieve stress? (Seriously, you don't? It was only six paragraphs ago. Damn this illiterate generation! Do you want me to tweet this book to you 140 characters at a time?) Our emotional state has a tremendous impact on our physical health, and appreciating humor, or more importantly the laughter it elicits, has been shown to lower blood pressure,[9] reduce blood sugar in diabetics,[10] improve our immune system,[11] and even help reduce pain.[12] I will get to all of these later, but I think the first two points merit repeating. Humor reduces blood pressure and blood sugar. With blood pressure, laughter assists with vascular function, causing our blood vessels to relax and expand while at the same time reducing the hormone epinephrine (you might know it by is common name, adrenaline), which contributes to elevating blood pressure. As for blood sugar, it's been shown that simply watching comedic films (literally getting a daily dose of comedy) can improve post-meal glucose levels in people with Type 2 diabetes.[13] Because of these two effects, humor can help to offset poor lifestyle choices. I'll use myself as an example.

I mentioned earlier that I am a bit overweight (does the phrase "white, fat, male comedian" ring a bell?), and I always have been. Despite that fact, I've always been in really good health. Keep in mind, I'm not running, like ever, and I sometimes get winded putting on pants, but what I'm getting at is that given my size, someone might predict that I would have developed hypertension or diabetes by now. Full disclosure, I've had spikes in my blood pressure, nothing chronic, but the sugar has been fine. I'm aware that if I don't make some serious changes to my lifestyle, I have an increased risk of developing these conditions. But one of the reasons I haven't yet is related to my sense of humor. Stress is also a factor in both of these conditions, and my ability to laugh at things, cope effectively, and recognize that some (most) things aren't worth worrying about has helped me in ways I'll never be able to quantify.

Fuller disclosure because I know my mother is likely to read this (already sold one copy! Yes, mom has to pay), I am actively working to modify my eating and exercising habits. As of this writing, I've lost the same pound fifty times. Change is hard. I should probably stop celebrating with ice cream.

The analgesic properties of humor are also noteworthy; laughter reduces our sensitivity to pain, *physical* pain. It's been shown that laughter reduces our need for anesthesia in surgery or painkillers in recovery.[14] And that, well, that's really cool.

Humor isn't just important to overcoming certain conditions, like depression or high blood pressure—it can be preventative too.

Let's get back to the subject of emotional disorders. One thing I'd like you to understand about emotional disorders is that they are largely behavioral in origin. What I mean by that is that the behavioral (or cognitive or emotional) choices that we make throughout our lives contribute to the development of the disorder. To put it simply, it isn't the case that you worry so much because you have an anxiety disorder, but you have an anxiety disorder because you worry so much. You can literally worry yourself into having a mental illness. Indian guru Meher Baba and later singer/human musical instrument Bobby McFerrin were onto something with the expression

"Don't worry, be happy." Excessive worry is just one of the poor choices we make that contributes to having a disorder. There are lots of other behaviors that help us develop these disorders—in other words lots of really bad choices. Even for those of us who may have some genetic or other biological predisposition toward having a disorder, it's ultimately the choices we make with our lives, our behaviors, which lead to these illnesses. Simply put, we choose to have them. It's analogous to body weight. I'm a "man of mass" if you will, a so-called "guy of girth." I was a "husky" sized kid, if you know what I mean. I am big boned, portly, corpulent, heavyset, chunky, big. . . . Let's face it, I'm fat (although I'm considered significantly underweight in Wisconsin). I didn't start out this way; it's the end result of a lifetime of poor choices. Some of them were behavioral, specifically those related to eating and (lack of) exercise, but they were also cognitive (I was taught to always clean my plate and never waste food) and emotional (eating is social and fun). There are some overweight members of my extended family, so maybe I inherited some predisposition to weight gain; but even then, it was the choices I just described that made my body what it is. My condition, although a biological one, is a product of my bad decisionmaking. Even in image-conscious Hollywood, a casual glance around town shows that I'm not alone in this, so I'm sure you can understand where I'm going.

Over the past fifty years or so, obesity has been rising. Americans in particular seem to be really good at gaining weight. There are literally tons of factors factoring into our accumulated tons, but much of it has to do with making poor choices more frequently. We are increasingly less and less physically active and consuming more and more. Like with obesity, more and more of us are making poor choices regarding our emotional health all the time. For some reason, emotional disorders are very popular. Nearly a fifth of our population will suffer from an anxiety disorder at some point during our lives,[15] and 15 percent of us will suffer from depression.[16] For such horrible disorders, that is really disturbing to me. The fact that many of us could have simply chosen better for ourselves makes it more so. And these disorders have become increasingly popular over the past fifty years.[17] More and more of us are making bad choices all the time.

Our current treatments of depression aren't very good. We treat depression with medication and therapy. Medication works, but not for everyone, and using it to relieve the symptoms of depression comes with a host of potential side effects. For severe cases, medication is definitely a necessity, but drugs alone are not going to eliminate depression because depression is a function of behavior. Behavior change is key and therapy is a means to effectively accomplish this. Without changing our behavior, we risk relapse and are likely to be dependent on our medication for a long time, possibly for the rest of our lives.

Even the best treatments have painfully low effectiveness rates.[18] The problem lies with the human brain; it's pretty resistant to change. Once our brain learns a behavioral pattern or manner of thinking or emotional response, it is relatively impossible to eliminate it from our head. Go ahead, try to unlearn worrying. While you are at it, quit smoking, eating carbohydrates, getting stressed in traffic, and sitting on the couch doing nothing when you're at home. Now that you've successfully eliminated those habits from your brain, please teach me how to do it as well and sit back in one of your mansions and enjoy your giant piles of money.

Change is extremely difficult for us and there are psychologists and other professionals who are currently devoting their careers to helping people do the impossible. I would tip my hat to them if I was wearing a hat and knew what that phrase meant (it sounds good); they are doing work that I do not have the patience to attempt. BUT, and here's a thought, what if we prevented the depression from developing in the first place? What if instead of dealing with a problem after it fully develops, we give people the tools to keep themselves from having the problem? What if instead of trying to go on a diet now, I'd learned to better manage my eating habits earlier in life?

Maybe we haven't yet developed depression, but we tend to focus too much on the negativity in life. Maybe we haven't yet developed an anxiety disorder, but we worry too much about things that are out of our control. Maybe we can adopt healthier behaviors before our behaviors have a severe negative impact on our life. Humor is so important in the prevention of disorder. It is a natural tool to help relieve stress, overcome tragedy, cope with disappointment, elevate mood, encourage optimism, and create social bonds. All are necessary if we want to prevent the onset of depression. Coincidence? Not in the slightest. Let's learn to take life less seriously, laugh more, and enjoy ourselves. Our emotional and psychical health may substantially benefit by adopting better behaviors early on.

Now, as much as I believe we should all develop a good set of behaviors early in life to prevent developing some terrible condition later in life, I am not a major shareholder of Google so clearly I do not own a time machine. If we made the mistake of being raised without hindsight, we still have the opportunity to implement some of the wisdom I will share later to overcome current issues. Yes, behavioral change is difficult. Yes, we are prone to repeating mistakes. Yes, I will have another doughnut (some forms of change are harder than others). Everything I mentioned earlier can, through serious effort, help improve our emotional health. Changing our behavior is difficult, not impossible.

Is this just some fad?

I feel like there has been a lot of discussion on the health benefits of humor in the media lately. It could just be my selective attention (it also seems as if every comedian I started with has gotten their own TV show); however, I just ran a Google search for "health benefits of humor" and got more than fifteen million results, and that is way more than existed on the Internet fifty years ago, so I feel safe in my statement. Whenever a topic gets popular, particularly a health one, some of us might perceive it as being the next big thing, something that will grab our collective attention but that we will ultimately move on from. Diets come and go, wacky (and some not so wacky) therapies catch on only to be discredited later, and if my Facebook feed is any indication (note: it's not) people do seem really enthusiastic about this whole "healthy humor" thing.

On occasion, when people find out I am a practicing comedian with a doctorate, they've asked me if I've met Dr. Hunter "Patch" Adams, the inspiration for the 1998 movie *Patch Adams* starring the late Robin Williams. I haven't, but I was lucky enough to meet Robin a few times. In the movie, Patch is portrayed as a funny doctor, using comedy to help patients recover. Although the real Patch Adams has pioneered the use of humor and clowning in an otherwise boring, sterile environment, the benefits of humor have long been recognized.

The Bible contains a passage that reads: "A joyful heart is good medicine, but a crushed spirit dries up the bones" (Proverbs 17:22). I'm no doctor, but dry bones sound kind of bad to me. Later, in the fourteenth century, Henri de Mondeville, the "father of French surgery," recognized the benefit of humor in helping patients recover from . . . wait for it . . . surgery. He even wrote, "Let the surgeon take care to regulate the whole regimen of the patient's life for joy and happiness, allowing his relatives and special friends to cheer him and by having someone tell him jokes." That's right, folks, tell the poor guy some jokes! Man, I bet my material on leeches would have killed in the fourteenth century.

Later still, in the sixteenth century, English scholar Robert Burton used humor as a cure for melancholy and Martin Luther utilized a form of humor therapy when counseling people with depression. Luther advised against social isolation and encouraged patients to surround themselves with friends who make them laugh. Leech jokes anyone? Seriously, I have like fifteen minutes' worth. In the seventeenth century, sociologist Herbert

Spencer theorized that we could use humor to reduce tension. Even depressive German philosopher Immanuel Kant suggested that we use humor to restore equilibrium, although had he been more successful at being optimistic, we would remember him as "Immanuel Kan." In the 1930s, clowns were, for the first time, brought into hospitals in the US to cheer up children with polio,[19] setting the stage for Dr. Adams and all the doctors on *Scrubs*. (My editor just informed me that despite my understanding, the show *Scrubs* was not a documentary. Still I'm leaving it in.)

Then THIS happened. In 1990, Dr. Martin Seligman became president of the American Psychological Association and reprioritized the aims of psychology. Until that point, psychology had been overly focused on curing mental illness, but now we were to focus more on making people's lives more productive and fulfilling, and not just treat their disorders. The result of this move was the birth of the subfield positive psychology, the study of positive human functioning. In my opinion, one of the best things to come out of positive psychology is our increased interest in positive affect and studying the things that make us happy. Like humor.

If the modern emphasis on the health benefits of humor seems like a fad, it might be because of our fairly new focus on positivity. As we've seen with some of the historical examples I gave you earlier, humor and its usefulness are not new.

So now you understand a little about how humor can improve our health, how it can help prevent certain types of disorders, and that we've known about this for centuries. But what is humor? What makes us laugh? How can humor aid in recovering from many different kinds of illnesses? What can we do that will make us laugh?

Well, you're just going to have to keep reading to see that "laughter really is the best medicine."

WHAT ARE WE TALKING ABOUT, ANYWAY?

B EFORE I GET INTO THE BENEFITS OF HUMOR AND LAUGHTER, I'D like to discuss what these things are. Unless you are a psychologist or a sentient robot trying to understand humanity (and sometimes those are not mutually exclusive categories), you may not have given much thought to the phenomenon of humor. Most people laugh and enjoy humor on a regular basis without ever questioning why or what humor is. Even those of us who produce humor on a regular basis, who seemingly have an inherent understanding of how to make others laugh, don't question it often. I've been making people laugh my entire life and it hadn't really crossed my mind until not that long ago to consider what humor actually is. Funny people rarely analyze their humor; it's just a part of their personality. We tend not to question what comes so natural to us and leave it to others to figure it out.

Robert Benchley, an American humorist, once said, "There seems to be no lengths to which humorless people will not go to analyze humor. It seems to worry them." In my experience, I have found this to be the case. People with limited responses to humor are always the ones who ask to have it explained to them, and with good reason: we want to understand the unknown. As I write this paragraph, I am sitting in a café in Hollywood, California. Within a ten-mile radius from where I am, I know of at least five studios that teach stand-up comedy to aspiring comedians, and without my having had experience with all of them, I can reasonably guess that if I were to take a survey, I'd find at least five definitions of humor. I would also find a lot of unfunny people hoping to learn how to be funny. The Los Angeles comedy scene is filled with unfunny actors whose agents have told them to give stand-up a shot (to be fair, casting calls are also filled with comedians who can't act hoping to score some acting work).

Humor is an art, and like most arts we know it when we see it but we might not be able to define it. In my seminars, I define humor as the tendency of certain experiences to elicit laughter or amusement. This is about the best definition I've encountered, although it doesn't quite satisfy my taste for something specific. It's a broad, ambiguous definition, and it has to be because there are so many things that have to be accounted for. Basically, humor is anything that we find amusing. The definition isn't limited to stimuli that provoke laughter, because there are plenty of occasions when we are amused but not laughing out loud, but it does need to include a wide variety of experiences. Humor can be experienced by hearing a joke, reading a funny comic, faking out our dog in a game of fetch, wearing a stupid outfit, seeing someone slip and fall, or even being tickled. Virtually any type of experience can result in a perception of humor under the right conditions, as can any sensory modality.

That's right; all of our five senses can be a potential source of humor. Psychologists sometimes recognize more than five senses, but I'll keep it simple and just describe those we are most familiar with: sound, sight, smell, taste, and touch. A lot of the humor we experience relies on our sense of hearing; for instance, we hear and tell jokes. This is verbal humor—someone says something that is perceived as funny. I include reading written humor in this category, although technically we aren't hearing an external stimulus, just our own internal voice. Maybe, hopefully, you've been amused at something you've read in this book so far, if not . . . well there is a reason comedy clubs insist on a two-drink minimum. The more you drink, the funnier we get.

Verbal communication isn't the only source of humor our ears provide us with; certain sounds strike us as humorous given the right context. This is auditory humor. Some sounds can be very funny. My favorite example of this is for you to imagine that you are in school, in some testing environment — let's say it's for some standardized test like the SAT or GRE, or GED even (ain't nothing wrong with those, and I've taken all three). I'll imagine it with you. These environments tend to be extremely serious; everyone is silent and focused on taking the exam. Obviously the results of these exams have some bearing on our futures, so we need to take them seriously; there are rules and time limits, so we stay as focused as we can. Everyone is silent and the only sounds are the scratching of pencils against Scantron sheets. Then suddenly, from the back of the room, we hear the familiar sound of somebody passing gas. That's right, somebody ripped a big old fart and it echoes throughout our testing center. Now, a lot of us are going to laugh at that. I personally think farts are hysterical and would likely lose my focus and laugh out loud in that situation. Imagine it's a particularly long fart, so acoustically perfect that you can almost picture the butt cheeks flapping against each other for a solid minute. At some point you can even hear a hint of moisture. . . . Okay, I may be getting a little graphic, but the point I'm making is that it is a funny sound. Farts are funny.

Humor can also be visual. We see things all the time that we find funny. Sometimes it's intended to be funny, like a cartoon or a performance, but a lot of the time it's just something that catches our eye and makes us laugh. The Internet is full of visual humor; Facebook and other social media sites are flooded with "memes" (basically pictures with captions added) and funny photos that are repeatedly passed around. I'm not a huge fan of the site, but if you've never visited www.peopleofwalmart.com

give it a peek some time for some examples of unintentional visual humor. By the way, if you see yourself on that site, I'm sorry ... but maybe this could be a first step to not taking things so seriously?

Obviously it's not just images that we find visually amusing, but also scenes and behavior. Now, what I'm going to describe to you isn't going to sound very funny in this format, but visually speaking it is. Just trust me. Imagine you are sitting on one side of a street, just hanging out people-watching. Across the street you see a nicely dressed man, three-piece suit and bow tie, walking down the sidewalk carrying a very large shopping bag overflowing with oranges. He's holding them in front of his body and trying to see around them as he navigates the sidewalk. Now imagine there's a banana peel on the sidewalk and he doesn't see it. You can already guess what happens next, but imagine he slips on that banana peel and his legs fly out from underneath him. He falls on his butt, throwing his bag of oranges into the air as a result. He is now on the ground as his oranges rain down upon him and he's desperately trying to catch them and recover the ones that are currently rolling all over the sidewalk. Now, I told you in advance it wouldn't sound very funny described to you, but if we were to see this scene play out in front of our eyes a lot of us would laugh at it. Not because we are horrible human beings with no sense of empathy, but because that shit is clearly funny.

What I just described might be called slapstick comedy, or physical comedy. It's a genre of comedy that appeals to something really basic in all of us. It's been around since Shakespeare and was very popular during the silent movie era. Charlie Chaplin and Buster Keaton were masters of this and their films still make me laugh. The Three Stooges, Laurel and Hardy, the Marx Brothers, the *Tom and Jerry* cartoons (and pretty much any other cartoon from that era, before society became more sensitive to cartoon violence) are all great examples of slapstick comedy from my childhood. Early in his career, Chevy

Chase was a master of the pratfall, a fall onto one's butt, which he demonstrated as a cast member on *Saturday Night Live*. Modern comedy movies still include quite a bit of physical comedy.

Physical comedy is not to be confused with physical humor. Physical humor involves our sense of touch. Sometimes, under the right conditions, certain forms of touch can elicit a humor response. We call this tickling.

Hearing, seeing, and touching (or feeling) probably account for the vast majority of our humorous experiences; but as I stated previously, any sensory modality could be a source of humor. Our sense of smell can make us laugh; our sense of taste can provoke amusement. Regarding smell, imagine we are back in that testing center all serious and focused. Imagine that instead of ripping a loud one, the farter produces the dreaded "silent but deadly." Now, the smell has to hit our nostrils before we know what happened, but once again some of us might laugh as a result of our sense of smell. Taste is a little harder to find examples of, but thanks to a recent experience I do have one for you.

To give this example, let me first go off on a tangent and give you a little back story. I'm originally from the Northeast, Long Island, New York, to be more specific. The Northeast is littered with these restaurants we call "diners." Now I realize the rest of the country has places they call diners too, but not like back east. Diners are restaurants that offer a ridiculous variety of menu options. Seriously, some diner menus have more pages than a book. How could one establishment offer an entire Greek menu, and have separate pages for Italian, kosher, Mexican, and everything else you could possibly imagine? Forget training someone on how to prepare all the different foods, how do they even have room for all those ingredients? Anyway, I've always been amazed by diners. A lot of them operate around the clock and offer Long Island teenagers, such as the one I used to be, a place to hang out and meet friends. Sometimes I went there to socialize, sometimes on dates, sometimes to study, but I spent a lot of time in diners. There was a practical joke that I remember some of us would play at the diner. You might be familiar with it already; we'd take the tops off the sugar and salt shakers and pour the salt into the sugar without mixing it up. Are you following me? When you put the tops back on them, the sugar shaker now has an undetectable layer of salt on top of it. We'd set that practical joke in motion, but I don't remember if I ever witnessed someone falling for it.

That was years ago and I'm no longer a teenager nor live in the Northeast, but recently I was on tour in New Jersey and I popped into a diner for a late-night coffee. When I got my coffee, I poured in my milk and added a few shakes of sugar. I stirred it up and took a sip, and it was the most disgusting, saltiest mouthful of coffee I have ever tasted. I spit it out and immediately started laughing as I realized what had happened. My sense of taste, in the context of my memory, provided me with a humorous experience.

Not to complicate things but because my seminars tend to appeal to a somewhat professional, educated crowd, people often ask me what the difference is between "wit" and humor. The difference is simple, humor makes us laugh. That previous statement is witty. To give you a little more, wit is clever. It can be subtle and usually involves a degree of thought, but it's not always funny. Witty remarks can make us laugh, but often we react with a simple "I get it" response. Wit is fantastic mental exercise and some of it can be really funny, but I think I prefer a good gut-busting laugh any day.

Anything can potentially be a source of humor. It isn't limited to verbal communication, and that's one of the greatest things about it. We are capable of laughing at virtually anything, provided it's funny.

SO WHAT MAKES
SOME THINGS FUNNY?

IN THE PREVIOUS CHAPTER I INTRODUCED THE IDEA THAT ANY-thing could potentially be funny. But why is the sound of the fart so hysterical in the testing setting, but not the grocery store? Why is it that when we see our loved ones slip and fall it is far less amusing to us than when Chevy Chase does it? Why would most people, I'm assuming, react to a mouthful of salty coffee with disgust and anger instead of laughing their asses off?

We may not have a precise definition of what humor is, but we do have a really good idea of why things are funny. There are several prevailing theories of humor out there, and each of them has their proponents. But my favorite theory of humor is called benign violation theory (BVT)[20] and was developed by Peter McGraw and Caleb Warren. Before I explain the theory, I want you to know why it's my favorite. It's not because McGraw and Warren owe me money or anything (they don't), but because in my humble opinion this theory has two things going for it. First, it is consistent with how the brain processes information, and second, it pretty much explains EVERY INSTANCE OF HUMOR IMAGINABLE. Seriously, I cannot think of a single example of humor that is not explained by BVT, and I've tried. It's an extremely robust theory and it's consistent with what we know about the brain's response to humor, so I put a lot of value on BVT. Also, I'm too lazy to come up with something better. Theories are hard.

According to benign violation theory, humor occurs when the following three conditions are satisfied by a situation. First, we have an idea of how things should be or we make a prediction about what is going to occur. Second, what actually occurs is inconsistent with our ideas or predictions; in other words, it violates our original appraisal or expectations and it does so in a nonthreatening way, and so is a benign violation of our expected experiences. Third, we are in a position, cognitively speaking, to recognize the difference. BVT says that for the experience to be humorous, the violation must be something that could potentially be negative in another context. And timing is a factor, as the threat appraisal and benign reappraisal of the situation should occur close in time. Basically, BVT explains that humor occurs when our brain recognizes it made a mistake in judging a potential threat. To anthropomorphize our brain a little, it's as if it is saying to itself, "I thought this one thing was going to happen, but something else happened and it's not scary or threatening. It's okay, I can laugh at it." Our perception of humor is the result of the discrepancy between a negative prediction and a benign outcome.

As I mentioned, I feel this theory can potentially explain anything we find humorous. Remember that fart sound I described early? Sure you do. Man, if only this book was printed in scratch and sniff. Well, anyway, that fart elicits laughter from us because it fits the theory. In that quiet testing center, we are focused and we have an expectation that the room is going to remain silent for the duration of the test. Our expectation is shattered when we hear that sound; someone literally broke the silence by breaking wind. It's a violation of our expectation and it's nonthreatening, it doesn't hurt us. Well, it's nonthreatening for those of us sitting far away from the source, but you get the idea. It violates our implicit expectation of a silent testing center and our brain reacts to this violation by perceiving the humor. When I asked Dr. McGraw to review an earlier version of this chapter, he pointed out that people are simply "not supposed to fart in public" and that "it is a bit disgusting." Which is the original source of the violation, in any context.

Think about the man walking down the sidewalk carrying his bag of oranges. When we see the man walking, our brain makes the assumption that he is going to continue walking down the sidewalk. Suddenly he slips and falls and is showered with oranges. What actually happened violated our idea of what was going to happen, and it was nonthreatening. It didn't happen to us. It didn't happen to anyone we cared about. It happened to an unknown stranger who reacted in the most spectacularly absurd way possible and we were in a position to recognize the discrepancy. Seriously, we had front-row seats. Think about the benign nature of this situation for a moment: it isn't us or anyone we care about. If it was our spouse or maybe one of our children that just fell down before our eyes we'd probably react a little differently. We would react with concern, we might be worried, and we'd probably run over to make sure they were not seriously hurt. THEN we'd laugh. Filmmaker Mel Brooks articulated this very well when he said, "Tragedy is when I cut my finger. Comedy is when you fall into an open sewer and die."

If we are overly empathetic, we might not find the situation funny. We might find ourselves thinking something like, *That's not funny, what if that happened to me?* Although we aren't the man who slipped, we might imagine what it would be like if we were in his shoes and react as if the violation IS threatening. Sometimes empathy gets in the way of a good laugh and I suspect this is one reason why men tend to enjoy *The Three Stooges* and the slapstick genre in general more than women. However, before you think of us as heartless animals devoid of empathy, show any guy the zipper scene from the movie *There's Something About Mary* and watch him squirm in empathetic agony. I can barely type this sentence as I've just conjured up that image in my own head. If you haven't seen the movie, you should cuz it's awesome. The scene involves Ben Stiller getting his testicles caught in his zipper in the most . . . oh man I just can't finish that sentence. It hurts just thinking about it.

I think you probably get it by now, but I want to make sure I beat this horse until it is good and dead, so just in case let's look at my sugar and salt practical joke. Recall that this joke basically results in a poor, unfortunate sugar-loving sucker getting a mouthful of salt. The expectation is obvious: we pour sugar, we expect sweet. The violation is the saltiness. How benign this is depends on our individual sensitivity, perception, past experiences, knowledge, and affective state. When it happened to me, I thought it was hysterical, but I immediately knew the joke and remembered setting it up for others. Plus I probably didn't pour too much salt in my coffee (I normally don't use sweetener at all, but this was a late night and I was hoping for the added pick-me-up), and I'm generally a pretty happy person which influences my reactions to things. Put all that together and I interpreted the violation to my expectation as nonthreatening, just a dumb joke. Alter any of those variables and I probably wouldn't have laughed. If I was in a bad mood, feeling anxious or depressed, I might have reacted as if it was just one more piece of evidence that the world is out to get me. If I had poured several heaping spoons full of the stuff, like thirty, it would have been so severely salty that my disgust would have been too great to overcome. Obviously I wouldn't have laughed if I had no prior knowledge of or experience with the gag. I can imagine an alternate, grumpy me being really pissed off and demanding a free meal from the diner for this horrible experience. Is that horse dead yet? I hope so. I think it's the one that put salt in my coffee. Horses like salt, right?

Imagine if my sugar and salt practical joke was reversed. What if instead I was at that diner and I ordered a plate of French fries, reached for the salt, and proceeded to unexpectedly sprinkle them with sugar. Even if I was unfamiliar with the joke, that seems far less likely to negatively affect me. So I now have sweetened fries; worse things have happened.

It is noteworthy that the violation must be something that would otherwise be perceived as a negative, or some sort of threat, for it to provoke a humor response. One day I was sitting at home working, with very few expectations for my day, and I suddenly got an unexpected phone call from a man offering me an amazing job opportunity. That didn't make me laugh at all. Neither would randomly bumping into an old friend that I haven't seen in ten years at the Baltimore Washington Airport of all places and discovering that both of us now live in Los Angeles (true story by the way). Unexpected positive events make us happy, but we rarely perceive them as humorous. Any comedian will tell you, it is very difficult to make positives funny, unless we find some negative spin on it.

I hinted that I recently ran into an old friend. It was my friend Kunle, an interesting physical character, muscular and nearly seven feet tall. He is the kind of person you instantly recognize when you see him. We first met when I was a graduate student and he was in college and were really close friends despite losing touch after our respective graduations. About ten years ago, I was standing at a hotel bar in Seattle waiting on the bartender to give me some attention and I noticed a giant black man standing next to me. He looked familiar, and I asked him if his name was Kunle, and it was him. Neither of us were staying in the hotel. I had popped in so my friend could use the bathroom and he was waiting to meet some colleagues. Random chance brought us together. We took a few photos, exchanged email addresses, and then promptly lost touch again, as he had to run off to join his coworkers, accidently leaving me to cover his bar tab. Fast-forward to this past May, I was on a seminar tour in the DC metro area and flew into Baltimore Washington International. Literally as I was picking up my suitcase at baggage claim, I heard someone shout, "Is that BRIAN KING?" and I turned around to see my friend Kunle, who had not changed at all. We were both on the run, but again managed to take a few photos and exchange information. I suggested that we try to get together and said, "You still owe me for that drink." He said, "I'll get you next time."

Okay, I tacked on the end of that story trying to find a negative spin to my very positive and unexpected reunion with my old friend. I'm not sure if it reads funny or not. If it does, great and you are welcome. If not, it just supports my assertion that it is very difficult to make positive experiences funny.

One more thing before I'll leave this alone: All three conditions I listed earlier must be present in the experience for us to perceive it as humorous. If we take any away, the remaining experience is not likely to provoke amusement. Let's say we don't have an expectation about the situation; something happens and we react to it. That's not very funny. That's basically the majority of life. Suppose we have an expectation that is not contradicted or violated. That's not likely to be funny either, because we expected something to happen and it happened just like we thought it would. Have you ever seen the TV show *Jackass* (or any of the movies)? It's a show that aired on MTV from 2000–2002 and to date has spawned at least four movies. It features clips of a group of guys doing painful stunts and playing pranks on each other and clearly has a wide audience. I don't find it very funny—sure there are moments that make me laugh, but overall it doesn't do it for me. Here's a hypothetical *Jackass* scene: Star Johnny Knoxville introduces the stunt by telling the camera, "This one is called Taserin' the testicles!" Then, they pick another cast member to willingly allow someone to shoot a Taser at his testicles and then THEY TASER HIS TESTICLES! He screams in pain or wets himself or both. For me, there is no violation of the expectation. When you Taser someone's testicles, my brain assumes that person will react with discomfort, and when they do, I don't find it funny. That's okay, I'm sure the guys from *Jackass* aren't up at night in their mansions stuffed with money and supermodels losing sleep because an unknown comedian doesn't get them. But for us to find something amusing, an expectation needs to be violated. Finally, we need to be in a position to recognize that our expectation has been violated. Without the ability to cognitively appreciate the difference, we are not likely to perceive humor. If we have diminished capacity or are not paying close enough attention, we aren't going to register the result as humor. Try watching a comedy in a dark theater where your attention is focused on the large screen in front of you. Now try watching a comedy on your TV at home, while simultaneously surfing the Internet on your laptop or smartphone. Notice a difference? Okay, I know you didn't do that just now, but I'm sure you can imagine that you'd probably laugh more in the theater than while multitasking. Our brain needs to recognize the discrepancy in order to process the humor. This also explains why drunks make bad audience members (putting a limit on my "the more you drink, the funnier we are" comment from earlier) and one reason why I envy musicians. (Also, musicians have groupies. Who among us wouldn't appreciate a few more groupies?)

I mentioned that I like BVT because it explains every example of humor I can think of and is consistent with how the brain works. I know I haven't gone into any detail on the brain yet, but don't worry, it is coming.

HUMOR: SO WHAT'S THE POINT?

O BVIOUSLY I'VE GONE INTO THE PREVIOUS DISCUSSION TO INTRO-
duce you to a theory of why we respond with amusement
to certain forms of stimulation. However, the main reason I
want you to understand this is to help reduce the negative emotions we experience. I believe that if we have a greater appreciation for why we laugh and what makes other people laugh, it
will serve as a first step toward minimizing the embarrassment,
distress, anger, and hurt we sometimes feel. Even if we personally don't think a situation is funny, we can recognize why people laugh, and when it is appropriate to laugh. Imagine if we
fall down in front of a crowd of people and they start to laugh
at us. Do we need to feel humiliated or can we appreciate the
humor of the situation? Maybe we are hurt and we don't feel
like laughing, but can we blame the onlookers for their natural
response to our misfortune? Understanding why people laugh
is a first step toward protecting ourselves from the pain of being
laughed at. It might also relieve us from guilt or distress we may
feel if our gut reaction is to laugh at someone else's misery. I'm
not saying we should have the freedom to ridicule others, but I
am saying that we should allow ourselves the freedom to laugh
when things are funny.

This discussion may benefit those of you who, for some
reason, have not developed a good sense of humor. You may

not find something amusing, but you can learn to recognize the source of laughter, the cause, and feel better about social situations regarding laughter. Maybe you'd like to blend into a group, and share a common experience without questioning yourself. I'm looking at you, Mr. Asperger's. . . .

Where does that sense of humor come from anyway, you ask? There are a lot of theories on humor besides benign violation theory. I haven't discussed them yet because I think they are not actually theories of what is funny, but instead they are theories of why we recognize or use humor. For example, Freudian relief theorists explain that humor is a means to relieve tension. Our fears, behavioral desires, and perhaps social situations produce tension within us and humor helps reduce that tension, providing us with relief.

People often ask me where I got my sense of humor, and to be honest I don't know. I wasn't a psychologist when I was growing up, I was just, you know . . . growing up. I did not have the foresight to take notes on my own development, but I have some ideas. One potential source of my sense of humor may have been my desire to relieve tension. See, I have always been a big guy but I have never been a physical guy and when I was a kid I couldn't fight. Fights are common, especially if you insist on being a nerdy fat kid, and I learned very quickly how to talk my way out of them. I developed the skills to discourage aggressors by making them laugh. To date, I have talked my way out of many potential ass kickings. As Jay Leno once said, "You can't stay mad at someone who makes you laugh."

I spoke with producer, musician, and former correspondent on *The Daily Show*, Vance DeGeneres, who seemed to have a similar development. He told me:

Much like the phrase "music soothes the savage beast" I think humor is exactly the same. You can use humor just like you use music. You can use humor to defuse tense situations, and I know that's something, personally, that I've used since I was a kid. I used it when I was in school. When there were bullies, I found that if you are able to make a bully laugh, they left you alone. I think that translates from there all the way up through when you're negotiating with a foreign country. I think people who use humor, in any kind of situation, benefit greatly from that. It also shows that you're friendly. Like I said, it can really defuse a lot of situations that could end very badly.

I asked him if he could share a specific example of how he may have used his sense of humor to defuse a tense situation, and he recalled a terrific anecdote:

I was sent to interview the head of the Ku Klux Klan in southern Indiana for a story, and it was me and a producer for *The Daily Show*. We picked up a two-man crew in Cincinnati and drove two hours out to interview him.

We got a really, really tense and cold welcome when we got to the guy's house. He had the KKK sign out front with "white power" on the sign, a big KKK flag in his yard, guard dogs, and a guy came out with a gun in a holster and took our IDs before they'd let us in. It was really tense. Then they took us in the house, just me and my producer. There were about four or five of them, these big KKK guys, not a smile on their face, and then introduced us to the Grand Dragon. The first thing he said was, "We know who you are, and you're not going to make fun of the Klan."

They had found out that our show was not news, but a comedy show. They let us get all the way there, and they tell us to turn around and leave. It was the most intense situation I'd ever been in, in my life. It felt very threatening. I tried to use humor as much as I could to defuse the situation and to try to get them to do an interview. We probably spent fifteen minutes discussing it. I tried to lighten the situation as much as I could. It helped a little bit, but ultimately they said, "Get the hell out of here."

Over time, I also could defuse tense situations between other people, all by making a joke. This skill also helped me deal with authority, get out of trouble, and has provided more than a few opportunities. Comedian Gabriel Iglesias (you might know him as Fluffy) tells a great story about a time he was on his way to an Arizona gig from Los Angeles and got pulled over for speeding. He was actually pulled over along with another car containing two women who were driving the same speed. The police officer came over to the window to ask for his driver's license and recognized Gabriel from a Comedy Central special. Immediately, Fluffy and his friends started cracking a lot of jokes about the situation, entertaining the officer. The officer took Gabriel's information and then went back to talk to the women in the other car. Gabriel put the theme from the TV show *Cops* on his stereo and blasted it from his car as the cop walked, laughing the entire way. From his rearview mirror, Gabriel could see the officer write the women a ticket, and send them on their way as he came back to Gabriel's car. He said, "Okay, you are free to go," with no ticket. Gabriel asked, "Why did the girls get a ticket?" and the officer replied, "Because they couldn't make me laugh."

The police have a tough job. It must be full of tense, difficult situations, and I imagine that their dealings with the public they serve are rarely positive. But I find that a lot of them have a great sense of humor; it is part of their coping mechanism. It helps them relieve their tension too. Police officers, members of the military, and others with similar careers often practice something we call gallows humor. Gallows humor is humor about especially painful subjects and hopelessly stressful situations. When I did a ride along with the Denver Police Department recently, I was amazed at just how dark their jokes would go. I asked my friend John Ager about this. John is a comedian of twenty-two years and currently based in Reno, Nevada. Before turning to stand-up comedy, he had a very different career. He was a police officer in Pennsylvania, a US Deputy Marshall, and worked for the United States Air Force Security Police. To put it mildly, he has seen some bad shit and laughed at it. He said:

> First of all you tried to make fun, or light, of everything you could, because it was such a stressful job. I was in riots as an airbase ground defense, and when you had that many people coming at you, you tell everybody, "I've never seen that many middle fingers in my life," and usually it's just to relieve the stress that surrounds you with all the other people. As a police officer and as United States Marshall, it was just making fun of any situation we could. For example if we were transporting a prisoner and we would ask him, "Hey what do you want to eat," they would say, "Cheesesteak." You just start laughing. Of all the things in the world that you could have to eat that we would get them, they would go after a Cheesesteak.
>
> If you had a bad accident you'd go back to the station and you just look at each other and start talking and just find anything funny you could find that particular day or night, just to get rid of the sights that you saw and the things you heard.

I told John the Gabriel Iglesias story I shared with you. Although he personally would not have let him go on the ticket (so he says, but I imagine if Gabriel offered to take him on tour that might have changed things) he said, "I can understand it. You've had a stressful day, you're running people's licenses, you don't know what you're going to come upon, and then people start making jokes and it makes you laugh. That's a real tension reliever so I can't fault the guy."

I can also understand why that officer would let Gabriel go. It happened to me once too, although I was pulled over for a significantly less transgressive offense. I was driving to a gig in Reno, following the highway behind the Sierra Nevada Mountains through some very sparsely populated country, and I got pulled over. I wasn't speeding, I hadn't done anything like change lanes incorrectly or throw something from my car, so I was really confused by this. The officer came to my window and said he pulled me over because I had a pair of fuzzy dice hanging from my rearview mirror. I know he was serious, but there was no way I could take that situation seriously and my friend (also a comedian) and I started making jokes about how low the crime rate must be if they would bother pulling someone over for fuzzy dice. I should look into real estate there—don't worry, kids, this is the safest zip code in California! When he first pulled me over he said, "So you're a comedian? Got any cop jokes?" I didn't, but I damn sure do now. He let me go too, by the way, and now I think he follows my Twitter. He's the first to like any of my donut pics on Instagram.

So yeah, relieving tension may have contributed to my own development. Other people may develop their sense of humor a little differently, and humor might serve a different purpose for them. Some use humor as a defense mechanism. This means they use humor to express ideas or feelings that are too difficult to deal with in a more straightforward manner, kind of like gallows humor. They cope with painful experiences, memories, or situations by making a joke. By doing so, they counter the unpleasantness of their reality with pleasure. I don't think I developed my sense of humor in this way, but a lot of humor comes from pain.

If you watch enough stand-up comedy, you will eventually notice that comedians use whatever causes them pain as a source of material. Whatever negative situations they have dealt with or lived through, or whatever might make them different than other people, becomes the subject of their jokes. Short comedians talk about being short, tall comedians talk about being tall, fat comedians talk about being fat, and skinny comedians talk about being too skinny. Tall, fit, attractive comedians talk about . . . well, everyone else. Poor comedians talk about being broke, and comedians represent the entire spectrum of mental and physical disorders, all of which might end up as subjects of their jokes. . . . We call it "bringing their pain to the stage." For some of us, our sense of humor is the end product of dealing with pain—it is our defense mechanism. For example at a recent showcase I saw a comedian take the microphone and do a ten-minute set on how he just got divorced. The next comedian did a ten-minute set on the fact that he just got married. Remember, they take their PAIN to the stage!

You may have noticed that so far my discussion has focused on situations or experiences that we find humorous, but I haven't used jokes (other than practical ones) as examples yet. Jokes, as we will see, follow the exact same pattern of perception with regard to benign violations. But, before I get into that . . .

WHAT IS A JOKE?

"WHO HAS TWO THUMBS AND JUST USED THE SHOW *JACKASS* TO make a point relating to a modern scientific theory? This guy."

As of the moment I am writing this, that is my most recent Facebook status update. Like most comedians, I try to use social media to be funny (by the way, feel free to follow me if you are interested, www.facebook.com/drbriankingcomedy). Sometimes it works, sometimes it doesn't. With only sixteen likes and seven comments I wouldn't exactly call that one a success. But I would call it a joke. It doesn't sound like a joke, and it is technically the truth, so what qualifies this status as a joke? The intent. Jokes are actions that are carried out to intentionally inspire humor. Plenty of things make us laugh, not all of them are intentional, but jokes are crafted for that purpose. Some of them are a bit more obvious than my status update.

The sugar and salt practical joke I mentioned earlier might seem mean-spirited, especially to the recipient, but it was set in motion to be funny. One of the problems with practical jokes, or pranks, is that the outcome is often removed from the joker's intent and perception of humor is up to the victim. Actors George Clooney and Brad Pitt have a famous history of pranking each other. In May 2015,[21] Clooney talked about a joke he played on Pitt by putting a bumper sticker on his car in the shape of a marijuana leaf that said, "Fuck Cops." Just hearing that story makes me laugh; I imagine Brad Pitt getting pulled over for some minor traffic violation, sitting in his car with no knowledge of the message he's communicating to the police officer. It's the "Kick Me" sign from high school, only with a greater potential for a beat down. I love it, and I am sure that Brad probably found it amusing too. But what if it was on his grandmother's car? Or what if someone who was not likely to be recognized by the police was driving? I can imagine plenty of people who wouldn't have recognized that action as a joke. In fact, the Internet is full of prank videos gone wrong. Just because the intended audience doesn't get it, doesn't change its intention as humor. Jokes are misinterpreted all the time.

Jokes come in many forms. They can be complex, or represented by a single thought or gesture. Depending on culture, a rolling of the eyes, or circling of the index finger around the ear can be a joke. A tongue sticking out, middle finger raised, or a snap can be used to make a joke. A joke can even be inspired by a single word. Do you remember the movie *Wayne's World*? I'm just going to assume you do. It had a huge impact on our popular culture throughout the early nineties by giving us the word "not!" as a joke. As in, "I really like the movie *Wayne's World* . . . NOT!" It was painfully annoying, but it did make a joke out of a single word (provided it was tacked on to another statement). That single word negated the prior statement in the most obnoxious way possible. More recently, people used the word "seriously?" as a single word joke. As in "It's 2016 and you are still referencing *Wayne's World* . . . seriously?" Some would argue that these are not single-word jokes, but single-word punch lines. I'm okay with that, I've got more book to write.

My favorite joke ever is the famous one-liner by Henny Youngman, "Take my wife, please!" That's a great joke. Most of us have heard it before, but it still makes me laugh. It's so simple, clocking in at only four words, and yet so powerful. The reason this is my favorite joke is the simplicity. In four words

Henny manages to pack a pretty powerful punch. Good writing, in particular good joke writing, utilizes word economy (which is extremely difficult for a long-winded doctor prone to run-on sentences). The shortest number of words we can use to get to the laugh the better, and I don't think I've ever encountered another joke in four words or less that was as funny as "Take my wife, please." There is a beauty in its simplicity. And it still fits the benign violation theory. The first three words put an expectation in your head that he's about to start talking about his wife; he's introducing the subject. The fourth word contradicts that expectation in a nonthreatening way. We aren't his wife, and we know that he isn't really trying to give his wife away. I wish I could write jokes that good.

A lot of jokes follow similar formats that we are all probably familiar with. The question-and-answer format is pretty common, especially when we are younger. Basically, we ask a question and get an unexpected answer. It's a simple and easy formula that children seem to latch onto very well. "What do you call a boomerang that won't come back? A stick." "What is the largest pencil in the world? Pennsylvania!" "What did the carpet say to the floor? I've got you covered." "What did the floor say to the walls? I'll meet you in the corner." And so on. . . . These are TERRIBLE jokes, but you get the idea. Asking the question sets up the expectation for a reasonable answer and the stupid answer we get contradicts that expectation. Now an interesting thing happens once you establish this pattern of joke telling. The expectation is now for a ridiculous answer; therefore, NOT getting a ridiculous answer contradicts that expectation and the result again is humorous. This is why we have such jokes as "Why do fireman wear red suspenders? To hold their pants up." Instead of the jokey answer we are expecting, we get an obvious, truthful answer to the question and that in turn becomes the joke. Personally, I would say they wear red suspenders because red is FABULOUS, but same thing. This leads me to that damn chicken.

"Why did the chicken cross the road?" Duh, to get to the other side. This joke is OLD, with the first known printed version produced in *The Knickerbocker*, a New York monthly magazine, in 1847. We all know that joke and probably very few of us laugh at it, but the simple, matter-of-fact answer is actually pretty funny when you are expecting a ridiculous punch line. Over time the joke itself became the root of derivative jokes like "Why did the turkey cross the road? Because it was the chicken's day off." "Why did the frog cross the road? Because he was tied to the chicken." "Why didn't the skeleton cross the road? Because he didn't have the guts." These jokes are just awful, but hopefully you can see that at one point in time, and perhaps at one point in our own development, they provoked laughter. Now if you've never heard them before, please don't think they are mine. I haven't cited any authors of these because they are considered "street jokes." They are so commonly retold that their original sources have long been forgotten. Children keep these jokes alive and share them generation to generation. The same bad jokes we told to each other when we were younger are being told by the current crop of kids.

Question-and-answer jokes, knock knock jokes, and certain types of riddles are popular among children because their brains are still developing and their experiences are limited. Eventually we graduate into more complex joke-telling involving short stories. A short story joke will be a couple of sentences to maybe a paragraph long and will end in a punch line. They might even have very common stems, or set-ups. For example, if someone says to you, "I've got a story for you: a priest, a rabbi, and a nun . . ." you know you are about to hear a joke. No legitimate story starts with "a priest, a rabbi, and a nun" unless you are at Catholic Bible camp. And still, you are probably about to hear a joke. "A guy walks into a bar" is also a really common short story set-up.

A guy walks into a bar with a duck on his head. Literally, he walks in with a LIVE DUCK sitting on his head! The bartender says to him, "Hey man, you can't bring that in here!" and the duck replies, "I'm sorry, but I've got this guy stuck on my ass!" (Street joke, my telling.)

In 2002, psychologist Richard Wiseman published the results of his yearlong cross-cultural online research on jokes.[22] For a year, he hosted a website called LaughLab, where users rated the humor of jokes and submitted their own. When we conduct research on humor, we often need to make people laugh in experimental conditions. Therefore having stimuli that have already been shown to receive high humor ratings across various groups of people is an extremely valuable resource to psychologists. Prior to the LaughLab study, normative data on jokes was difficult to come by. The website attracted a lot of media attention, and received more than forty thousand joke submissions and 1.5 million ratings.[23] The jokes were in English, and at the conclusion of this monumental work, which remains the largest humor-related research project to date, Dr. Wiseman was able to determine the "World's Funniest Joke":

Two hunters are out in the woods when one of them collapses. He doesn't seem to be breathing and his eyes are glazed. The other guy takes out his phone and calls the emergency services. He gasps, "My friend is dead! What can I do?" The operator says, "Calm down, I can help. First, let's make sure he's dead." There is a silence, then a gunshot is heard. Back on the phone, the guy says, "OK, now what?"

The world's funniest joke was submitted by psychiatrist Gurpal Gosall, but the earliest telling of it may have been from comedian Spike Milligan in 1951.[24] It's a short story format joke that sets up an expectation and violates that expectation in a very funny way.

Speaking of the LaughLab, I find these points interesting given my previous discussion. First, among jokes that referenced animals, those that mention ducks were perceived as funnier than other jokes. Apparently ducks are the funniest animal, especially when a guy walks into a bar with one on his head. Also, the most frequently submitted joke was a question-and-answer–style joke: "What's brown and sticky? A stick." It was submitted more than three hundred times.[25] Clearly it is in the same vein as "Why'd the chicken cross the road?" and not a single participant thought it was funny. Perhaps poet Ralph Waldo Emerson said it best: "I dream of a better tomorrow, where chickens can cross the road and not be questioned about their motives."

Other common joke forms might make use of wordplay. Speaking of Facebook, I just took two Facebook quizzes: "Who's Your Soulmate?" and "Who's Your Cellmate?" Lucky for me, the same person was picked for both. You know who you are, big boy.

That's a little wordplay. It also includes some irony and a touch of sarcasm (spoiler alert, I am not a homosexual). As far as joke formats go, I love sarcasm. I'm a very sarcastic person and I use sarcasm a lot in my humor. Sarcasm also interests me as a psychologist, because it is a subtle form of humor that not everyone understands. With sarcasm, the humorous intent is conveyed not by the words that are used, but by the tone and context of the delivery, and there are people that just don't get sarcasm. And for some reason, the majority of those people chose to move to the Midwest. Now, if you don't get why that last statement is funny, you might be one of those people. That was dripping with sarcasm. Sarcasm is often used to mock or convey contempt, but it doesn't have to be mean spirited.

Hopefully, my statement about people in the Midwest is obvious sarcasm. Clearly there is no way I actually believe that an entire region of the country doesn't get a certain form of humor, but it could be considered a minor jab at the good people of the middle. It was partly based on an experience I had when I first moved to Ohio to complete my doctorate. I was in a small-town video store, just browsing at the shelves of VHS tapes (wow, that entire sentence just made me feel very old) and a clerk asked me if he could help me find anything. I was in the Disney film section and I said something like, "Yes, I'm having a hard time finding your porn." The clerk stared at me for a moment, startled, and eventually said, "Um, we don't carry that here." Now, consider the context. Was I really shopping for porn among Disney movies or was I joking? I'll never forget the look that guy gave me. For me, the key point to sarcasm is that it requires an understanding of the delivery context, and from a psychological perspective, that interests me because it requires a high level of social intuition and communication between the two hemispheres of the brain. I'll touch on that in a later chapter.

Did you know that a tree that was planted in Griffith Park in Los Angeles as a memorial to former Beatle George Harrison was killed in 2014 because it was infested by beetles?[26] That's some real world irony right there. Also, his former house was broken into by a band of traveling Wilburys.

Regardless of the specific style, traditional jokes all follow a basic structure of "set up, punch." The first part of the joke sets up the prediction; this is the "set up" or "premise." It can be a question, a statement, a story, the words "knock knock," or anything the humorist uses to put that expectation into your head. The "punch" or "punch line" shatters the expectation. Punch lines can be the answer to the question, the last line of the story, the last word, or anything that changes the expected outcome of the joke. Traditional jokes use punch lines, but if you pay attention to modern stand-up comedy, you'll find very few traditional jokes. Comedians in general have moved away from the use of traditional jokes. This is an observation journalist Warren St. John made in 2005 with his *New York Times* article "Seriously, the Joke Is Dead."[27] The joke is not dead, but most professional comedians do not tell traditional jokes. They tell stories, make observations, do act-outs, voices, impressions, and incorporate props or magic, all of which are actions intended to provoke amusement. In other words, jokes.

I can't tell you why comedians do not tell traditional-style jokes anymore, but I suspect it has something to do with the evolution of the art form. Yes, comedy is an art form just like any other (don't feel bad if you didn't think of it that way, I wish more comedians realized this too). And like other arts, the art of making people laugh evolves and it always has. Some of the language used and topics that comedians cover now was considered taboo just decades ago. In fact, comedian Lenny Bruce was arrested in 1961 for using the word "cocksucker" at a club in San Francisco.[28] Now, I'm not a big fan of profanity in comedy, but I hear way worse words being used by comedians all the time. Another reason I think comedians have moved away from traditional joke telling is to protect their material. All street jokes have to be written by someone, but they are so easy to retell and have been retold so often that many of their authors have been forgotten. If I write a traditional-style joke, anyone can tell it. But if I share my opinion, make a unique observation, or tell you a story about a personal experience, the joke and joke-teller become less separable. For example:

> Let me tell you a little about myself. I was recently eating breakfast, and I noticed right there on the container of cookie dough, the words "Please do not consume raw cookie dough."
>
> Now, do you skinny people know another way to eat cookie dough that I'm not aware of? You take a bucket of dough, a spoon, and some depression.... That's a meal!

That's part of a bit I tell on stage, and it is pretty unique to me. I suspect that the spoken version is much funnier than reading it, but this bit usually does really well. When I mention that I had been recently eating breakfast, it sets an expectation that what is about to follow is something traditionally eaten for breakfast like eggs, pancakes, or spaghetti. And I said cookie dough, which shatters that expectation in a humorous way. Then, I pretend as if, because I am overweight, I've never heard of another way to consume cookie dough other than eating it raw from the container. It's a non-traditional joke that includes a lot of my personality and is based on my unique experiences. Well, I have never actually eaten cookie dough for breakfast. Okay, there was this one time. And maybe a few more, but that was it. Okay, truthfully, if you ever see me perform live, please feel free to bring some cookie dough. It's good for any meal.

JOKES: AGAIN, WHAT'S THE POINT?

ONCE AGAIN, I'VE GONE INTO PRETTY LENGTHY DETAIL TO explain something that a lot of us implicitly "know" or what might sound like common sense. I've done this because I think it is extremely important to be able to identify jokes for what they are: actions intended to provoke humor. Jokes are an important form of communication. We may not get a joke, we may never think that it is funny, but if we learn to recognize when someone is joking, again intending to provoke a response that normally makes us happy, that will go a long way to reducing the distress, anger, and other negative emotions we may experience in our lives. Imagine someone says something to us that can be interpreted in several ways, and maybe we get offended by it because we failed to recognize the humorous intent. Being offended might make us angry, cause us to experience distress or hurt our feelings, and possibly contribute to our overall negative affect. All of that can be avoided if we learn to identify when people are joking. Granted, not all jokes are funny, but if we are consistently being told by multiple people in our lives that we "can't take a joke" or if we are quick to anger, we probably struggle with joke recognition. Also, we probably aren't as happy as we could be either. And by "we" I mean "you" because I can take a damn joke.

Another reason why I believe it is important to understand jokes is from the other perspective. Suppose we feel as if nobody "gets" us. Every time we make a joke, it is misinterpreted and angers, offends, or unintentionally pushes people away from us. And again, by "we" and "us" I mean you, because I have tons of friends. Actually, I've had plenty of jokes backfire on me. I've had people in my life come and go because of my sense of humor. I've lost friends over things I've said and although I have a decent sense of humor now, I've struggled with what is funny in the past. I have a comedian friend who does nothing but offend people on stage. I know him well, and I know this isn't his intent, but he hasn't yet learned how to balance an audience's sensibilities with his own or how to communicate his punch lines. He comes off as creepy and angry, two adjectives that tend to get in the way of his humor being well received. If the audience thinks you are a creep, then your creepy punch line isn't going to be recognized as a benign violation to their expectations. Comedians piss people off all the time, they are used to it, and some of them unfortunately seem to enjoy it. However, for the rest of us consistently pushing people away can lead us to a pretty dark place. The point is that we don't want to be the asshole everyone avoids, so either learn how to tell a joke better or recognize what you are doing wrong and cut it out. I think it's really important to our ability to get along in the world if we can properly identify jokes.

THE BRAIN

"I used to think that the brain was the most wonderful organ in my body. Then I realized who was telling me this."

— Emo Philips

IF YOU DON'T RECOGNIZE THE NAME EMO PHILIPS, I STRONGLY recommend you look him up, especially if you enjoy quirky, intelligent humor. I was first exposed to his work in 1987, when HBO aired his special *Live at the Hasty Pudding Theater* and he struck a chord with me. His humor wasn't relatable; he was tall, awkward, sported a bowl cut, and wore a strange outfit that seemed like it was pieced together from the costume department of a low-budget Shakespeare production. I never followed his career beyond that special, but many of his jokes, including the quoted one, stuck with me. I won't mention how old I was at the time (this is Hollywood after all), but it was before I even entertained the idea of pursuing my doctorate. Years later, when I was accepted into a neuroscience program, literally focusing my research on the brain, I'd repeatedly reference that joke.

The brain is the most wonderful organ in the body. It's the source of all our experiences, everything we know, everything we perceive. Our entire lives occur inside our skulls, and of course our perception of humor is a function of brain activity. I promise you, I am not going to get too technical in my discussion of how the brain processes humor, but it's important to understand a little. Plus, every time I misspell my name, the result is "brain," so you know, there's that. If you can bear with me a bit, I'll get funny again.

Let's first get some basic terms out of the way. The brain is an organ made up of cells called neurons (and other stuff, but I'm not going to get too complicated) that sit inside our skull, protected from the outside world by layers of tissue and bone. It receives information about the outside world from the stimulation of neurons that extend into virtually every part of our body; we collectively refer to these cells as nerves. Sensations on our skin, smells, visions, and sounds are stimuli that are detected by specialized cells that convert the various types of information into electrical signals that can then be transmitted to the brain. Therefore we might see a scene play out before our eyes or hear someone tell a joke and that information, whatever the format, is brought into our brain for processing. We'll refer to the information coming in as stimulation, and each unit as a stimulus. The part of the brain that does most of our complex processing is called the cortex; it is the outer wrinkly layer of the brain. Inside, there are other structures that process information and are involved in more important things like keeping us breathing and making sure we eat when we need to. The brain is also somewhat split down the middle, creating two nearly identical hemispheres, the left and right. I apologize if I've gone too basic, but I want us all to be on the same page.

First, let's look at what happens when we encounter a stimulus, any stimulus. One of the first things our brain has to do is determine if that stimulus is going to kill us or not. That's a pretty important step—it helps keep us alive. If our brain has any chance of survival, it has to learn how to identify potential threats, and do so quickly. Our brain needs to make an assessment of our situation in a hurry if we are going to react accordingly. Now, think about a decision you've made recently. It doesn't matter what it was or how important. Hypothetically, let's say the last time you made a decision about what to have for lunch. Yesterday, I ate at El Tepeyac, which is a pretty awesome and well-known Mexican restaurant in East Los Angeles. My brain factored in a tremendous amount of information as I read over the menu considering my options. In no specific order my brain considered the following pieces of information: There was my physiology—obviously I was hungry but how hungry was I and what nutrients was my body craving? I considered my past experiences with different recipes—which have I had before and how well did I like them? Did I feel like getting something I am already familiar with or trying something new? What about the individual ingredients? Did I have any allergies? Did I really want everything that was included? Have I had any negative reactions to cheese or beans? I considered my company—what were my friends getting? Would they judge me harshly if I ordered the giant "Manuel's Special" burrito and attempted the challenge? What about my long-term health goals, and how will this decision influence my likelihood of reaching them? Ultimately, my brain decided on a delicious regular-sized burrito which I enjoyed and moved on with my life. Someday I'll attempt that challenge and get my free T-shirt!

My point is that my brain labored over that decision-making process for a few minutes at least, and that isn't even including the decision to eat Mexican in the first place, or our choosing El Tepeyac as the place to get it. What I've just attempted to describe is conscious decision making, and I want you to understand that conscious decision making can be a very slow process, even for seemingly simple decisions. Our conscious mind can take forever to make a choice. If I was in a potentially threatening situation and I had to rely on my conscious mind to consider all the relevant information before reaching a decision, I'd be screwed. Thankfully most of our decisions are not made by our conscious mind.

When we encounter a stimulus, our brain unconsciously assesses whether it is threatening, and for those of you who care this is one of the primary functions of a subcortical area in our brain called the amygdala. If the amygdala determines that a stimulus represents a potential threat, it will then send signals to other parts of our brain to initiate physiological responses. Those responses will include things like an increase in our heart rate, our hair standing on end, our face becoming flushed, and our vision and hearing becoming more focused by eliminating peripheral stimulation. Our stomach also tenses up, which is why we might call this a "gut" reaction. These physiological reactions get our body ready for action, and we collectively might refer to this as arousal or our stress response. It's a response that prepares our "fight or flight" behaviors. The decision to respond and the communication of that decision to other organs happens nearly instantaneously. In fact, some of the physiological changes will take place before we have any conscious awareness of the stimulus. In other words, we can be exposed to a stimulus, determine it is threatening, and respond to it with a change in our physiological condition before we even know we've encountered something. Because these decisions are made so quickly, we may consider our reactions as being automatic. That's fast, and that all occurs outside of our conscious mind.

When I use the words "conscious" and "unconscious," I want you to understand what I'm referring to is whether we have awareness of a decision or not. These terms are pretty archaic—scientists and philosophers argue if they are even still appropriate[29]—but I find that they are concepts that are easy to understand. So to keep this simple, I'm not going to debate consciousness.

The part of our brain that we might call the home of our conscious mind is an area called the prefrontal cortex. It is the part of your brain that lies directly behind your forehead and eye sockets. It's called "prefrontal" because it is the front portion of the larger frontal lobe. It's the most recently evolved brain structure, and the last part of your brain to develop. In fact, it isn't completely finished until our early twenties, which explains why, in our teens, we thought that ridiculous hairstyle was a good idea. It is responsible for executive functions in the brain, including planning actions and decision making. As I've previously explained, it's a much slower decision maker than our unconscious mind, but that's because it has the potential to incorporate an almost limitless amount of information into the process. That is conscious decision making, like what to have for lunch or where to go to college. It is involved in language processing and speech production and is the home of our inner voice.[30] It's the source of our awareness. Basically, it contains all of the activity we associate with thinking.

Sometimes our unconscious mind overreacts to our environment or misinterprets a stimulus as threatening when it really poses no real threat to us. One of the great things about the structure of our brain is that we can form a quick reaction but then modify or suppress that reaction as we need. Our prefrontal cortex has the ability to override decisions made by our unconscious mind. With a little thought, we can consciously "decide" to veto an action, modify it, or respond in an entirely different manner. Ultimately, we have a great deal of potential to influence our gut responses.

Humor is assessed by our conscious mind. Recall that when I discussed benign violation theory, I mentioned how our brain recognizes a violation to our expectations as being nonthreatening, and therefore humorous. This processing occurs in our prefrontal cortex and involves both hemispheres of the brain. When we encounter something potentially humorous, we may have an initial gut reaction but our cortex gets to work very quickly. We might consciously decide that our gut reaction was appropriate or we can recognize the mistake and adjust our reaction with laughter. With regard to jokes, the two hemispheres contribute to their interpretation differently. Although the left and right hemispheres are mostly similar, language functions are usually located on our dominant side, meaning that for 97 percent of us, our language functions are primarily located in the left hemisphere (we're right-handed). That means that our dominant hemisphere, the left for most of us, analyzes the structure and words of the joke. It knows the rules of grammar and has access to our vocabulary. It comprehends the literal interpretation of what has been communicated to us. Conversely the other side, the right for most of us, helps evaluate the intent of the communication. It factors in the context, the emotional tone, and other related information we may have on the subject. It allows the brain to understand the intent of the communication. In other words, the right hemisphere "gets" the joke.[31] Without the contribution of the right hemisphere, all jokes would be taken at their face value with their literal meaning. The right hemisphere therefore allows us to distinguish between a lie and a joke.

Most jokes are technically lies at face value. Earlier I shared with you a joke that I tell regarding eating cookie dough for breakfast. Every time I tell that joke to an audience, I am sharing something about myself that is not true. However, I have yet to encounter a single heckler to stand up and say, "Hey Brian, why are you lying to us? You didn't really eat cookie dough for breakfast recently, did you?" Nope, that has never happened. Most of the time people get that I'm joking (sometimes the joke is too subtle, so I will comment that they are looking at me thinking, *Yeah, that's about right*, or that as I entered the stage, they were thinking, *This guy eats cookie dough for breakfast!*). We need the contribution of both hemispheres to properly identify jokes. It makes sense then that damage to the right hemisphere, impairments of the language center, and damage to the corpus callosum (a structure in the brain that connects the two halves[32] and allows them to communicate) are all associated with deficits in humor.[33]

This brings me back to why I find sarcasm so interesting, because it is true that some people just simply do not get it. Let me return to the example I gave you about browsing the video store in Ohio. I was standing in the Disney section and told the clerk I was looking for pornography. At face value, this was a lie but I had intended it to be humorous. His reaction indicated that he did not recognize the sarcasm in my statement. Interestingly, he also didn't react as if I had lied to him. Maybe his gut reaction was so overwhelming that he didn't have an opportunity to consciously evaluate what had been said before forming his behavioral response. Or maybe, he responded before his right hemisphere could contribute its piece in interpreting the true intent of what I had said. I don't know what might prevent some people from perceiving sarcasm, but I do find it very interesting.[34]

As we are processing a joke, other areas of our brain become involved. Our visual area creates images that might be associated with the joke. Subcortical areas in our brain that are part of the limbic system become active to make us feel happier. Finally, areas of our brain responsible for planning and executing movements will make us smile and laugh. With the initial sensory processing, threat assessment, prefrontal analysis, and all other areas involved in our humor response, hopefully you can appreciate that humor has an enormous impact on the activity of our brain. With everything I've just listed, the vast majority of our brain is somehow involved in processing and reacting to humor. Even a fart joke. Hey, who ripped one?

THE BRAIN: WHAT'S THE POINT?

MY PRIMARY REASON FOR GOING INTO THIS DETAIL OF THE BRAIN is so that you have a better understanding of why we do what we do, and the mechanisms that underlie our perception of humor. Also, having a basic understanding of these processes will facilitate later understanding of some of the benefits of humor.

In the previous paragraphs, I hinted at all the various parts of the brain involved in our response to humor. Humor inspires a lot of brain activity. This tremendous amount of activity in response to humor means that humor can be used as a quick assessment of brain health. If you can laugh appropriately and/or make others laugh, chances are really good that your brain is fine. Some brain damage or conditions can interrupt this process and result in humor deficits.[35] Damage to either hemisphere or their ability to communicate with one another can interfere with our perception of humor.[36] If someone suffers some trauma, a great way to quickly assess their health is to attempt to make them laugh.[37] Conditions like clinical depression or Asperger's syndrome[38] are also associated with deficits in humor. However, I feel it is important to stress that although humor can be an indication of brain health, if we don't respond to humor well, it doesn't mean our brain is broken. There are lots of reasons someone may not respond to humor that have nothing to do with the health of the brain. So if you are one of those people, please don't worry about it.

SPEAKING OF LAUGHTER...

EVERYONE KNOWS WHAT LAUGHTER IS. EVEN THE GRUMPIEST, nastiest jerk has at least seen other people smile and laugh. Laughter is universal; it is present in every human culture and we all do it the same way. That, right there, fascinates me. You can randomly pick people from anywhere on the planet and regardless of their language and cultural history, they will laugh the same way in response to humor. They might differ on what they find funny, but they will certainly laugh at it. Laughter is a culturally universal behavior. As Russia-born comedian Yakov Smirnoff once said, "Everybody laughs the same in every language because laughter is a universal connection." If you don't remember Yakov, he was really popular in the mid to late 1980s, starred in a handful of movies and TV shows, and was famous for his catchphrase "What a country!" He now owns a theater in Branson, Missouri where he continues to perform on a regular basis. Coincidentally, he has since gotten a master's degree in positive psychology, making him yet another comedian-psychologist hybrid. Maybe I should learn to speak Russian.

Recognizing that laughter is a cultural universal then supports the idea that laughter is innate. It is part of our genetic inheritance, something that comes in the brain as soon as we open the box. No assembly required. Regardless of who we are and where we are born, we have the potential to laugh. We aren't born laughing, but we are born with all the tools to laugh as soon as we can recognize us some humor.

Another point I would like to make about our innate propensity toward laughter is that human beings are not the only species with a sense of humor. Laughter and playful behavior, a relative to humor, are observed in lots of other animals. The easiest animals to identify it in are chimpanzees, because they are so morphologically similar to us. When a chimpanzee smiles, its smile looks like one of our smiles, and when a chimpanzee laughs, its laughter resembles our laughter a lot (although the sound is a little different).[39] Chimpanzees also applaud and jump around in response to humor.[40] Dogs also laugh.[41] As social creatures, dogs engage in play with one another, or other companions, to amuse each other. Conversely, cats do not laugh, but may have a sense of humor.[42] Cats seem to play just to amuse themselves, those selfish jerks. I am a dog person; I just love them, and one of the things I appreciate about dogs is how playful they are. Dogs are fun-loving creatures that don't take life too seriously, and they certainly seem to enjoy it when we laugh. They will engage in behaviors with the sole intention of making human beings laugh, and I love them for it. They also laugh. We can't hear their laughter because our ears can't register it but more sensitive instruments show that dogs laugh. They laugh when they are playing, and they laugh at us, a lot. Dogs must think human beings are ridiculous.

Other mammals also laugh but the further down the evolutionary ladder you go, the harder it is to recognize. Even rats laugh. Jaak Panksepp, a professor at my alma mater, Bowling Green State University, was one of the first to document this.[43] We cannot hear their laughter without help (good thing for me, with all those late night study sessions) and it appears to be mostly in response to play and tickling, not an appreciation of humor. Laughter does seem to be specific to mammals; there aren't any reptiles out there cracking up or telling jokes.

Have you ever wondered about what laughter is or why we do it? I'm guessing probably not—unless you are also a comedian or a psychologist that studies humor, like me and Yakov, you might not have given this basic human behavior much thought. Laughter is present in every human culture and likely predates the evolution of human beings, given its presence in other animals. Whatever it is, it is probably pretty important.

We live in an era of digital communication. I use my "phone" more for taking photos than taking calls. When we communicate via texting or chatting on the Internet, it's hard to convey emotional responses. Therefore, over time people have developed ways to emote via a form of communication that limits emotional expression. To convey the emotions we feel, we share photos, little graphics that represent various facial expressions called emoticons, and common acronyms such as "LOL." For those of you still on dial-up (hi, grandma!), "LOL" is an acronym that stands for "laughing out loud" and I'd be seriously surprised if I needed to tell you that. It is so commonly used, online and off, that "LOL" has been entered into the *Oxford English Dictionary*. Even my spell-check program recognizes it as a valid word. Lol, lol, lol. The only error in that previous sentence is the grammar. We write "LOL" because our audience cannot see or hear us laughing, and they cannot hear our tone.

I suspect that "LOL" actually, yet unofficially, stands for "I am deceiving you." I will let you in on a secret: nearly every time I type those three letters I am not actually laughing out loud. If you use it, I bet most of the time you aren't actually laughing either. I think it has sort of evolved into punctuation, and we use it more for that purpose now.[44] Just like the exclamation point is telling the reader that this sentence was written with a tone of urgency! Or the question mark tells us that we are unsure, maybe? I think "LOL" is mainly used to let our reader know that the previous statement was intended to be humorous. I have to include it with sarcastic statements all the time, especially when chatting with my friends from Ohio, lol.

When we use "LOL" as a response to someone else's text, we probably think that what we've read is amusing, but again I bet we are rarely actually laughing out loud. To convey that we are genuinely laughing, we have other acronyms that step up the game a bit. "LMAO" translates to "laughing my ass off," which, although hopefully not literal, probably does accompany actual laughter. Some people use the acronym "ROFL," which stands for "rolling on the floor laughing," to express the most extreme laughter reaction, and again I bet almost everyone using this acronym is lying. Nobody is currently rolling on the floor next to their computer with a hand free to type that out. At best, you are laughing out loud for real. (Maybe LOLFR could catch on?) Obviously the intent of using ROFL isn't to deceive, but to express the intense amusement we find in the text. It may be an exaggerated expression, but there is some truth to the idea of rolling on the floor in an uncontrollable fit of laughter.

Laughter is how our body responds to humor, and it can be anything from a few chuckles to a loud, uproarious sidesplitting guffaw. It starts with the face, with a smile, then that smiling turns into laughter. It involves the use of fifteen different facial muscles.[45] Something really peculiar happens when we begin laughing: our larynx, the little muscular organ in our throat that we sometimes call the voice box, half closes.[46] This tightening of our throat makes it very difficult for us to breathe and because we are now at risk of not getting enough of that sweet, sweet oxygen we seem to like so much, we begin gasping for air. That is where the sound of laughter comes from, literally the sound of laughter is simply us GASPING FOR AIR![47] If we continue, the struggle for oxygen will activate our tear ducts and we'll start tearing up, maybe even crying.[48] I've heard people say things like, "I don't know why I'm crying, it's just so funny!" Well I know why you are crying, you are crying because your brain isn't getting enough oxygen! So cry away, let those tears flow!

Struggling for oxygen reminds me of a quick anecdote. When I first started dating my girlfriend, Sarah, we took a trip to Colorado Springs and drove up Pike's Peak. Still in that hon-eymoon phase, we felt blissfully, smiley happy the whole time we were together. Seriously, we were obnoxious. Well, because I was really accustomed to living relatively close to sea level in Los Angeles, the altitude of the mountain really affected me. I began experiencing shortness of breath at around seven or eight thousand feet. Sarah was a little more acclimated and did not begin to feel it until later in the drive, but by the time we reached the summit she was also having difficulty breath-ing. The summit of Pike's Peak stands at an impressive 14,115 feet above sea level, enough that both of us were winded just getting out of the car to take a picture. Then the most weirdly wonderful thing occurred as we were looking out at the view that inspired the song "America the Beautiful": we both spon-taneously burst into a fit of uncontrollable laughter. It started with a giggle, a chuckle, and then there we were standing on top of a mountain laughing our asses off like a couple of goo-bers. This being Colorado, I should probably add that I swear we were not high on anything other than altitude. We had no idea why, but we simply could not stop laughing. Looking back at that moment, there was not really anything to laugh at but I suspect that our brains were significantly oxygen deprived and our laughter helped replenish our supply. Also, it didn't hurt that we were already full of positive emotion. It was a magical moment that neither of us is likely to forget.

With all that air gasping and mouth moving, clearly there are quite a few muscles that become active during laughter. Our facial muscles and our diaphragm get a nice workout when we laugh. If the laughter gets more intense, the rest of our body might join in the party, particularly the muscles of our limbs and back. We may start to have involuntary contractions in many of these muscles and our arms will just start flapping up and down. That sounds ridiculous, but think about it for a moment, what happens to our body as we laugh? We might raise our arms automatically. If so, what should we do with a set of newly raised arms? Let's bring them together! This is where clapping comes from, the muscles in our arms contract involuntarily and we make sense of the whole thing by turning it into a move. We are so smooth. Not all clapping is involuntary; some of it we do intentionally, but I bet everyone reading this has at some point laughed so hard that they were moved to applaud automatically.[49] Applause isn't the only way we can take charge of a pair of involuntarily raised arms. Imagine you are in the midst of a laughing fit and your arms spring up all of the sudden, well we have to show those suckers who is boss and bring them back down. Maybe we bring them down against the side of our leg, because, you know, we meant to do that all along. That's where knee slapping might come from. Clapping and knee slapping are just two common behavioral responses expressed thanks to the involuntary muscle contractions we experience with very intense laughter.

And it isn't just our arms—our leg muscles also contract. If we are standing, we might not notice it as we maintain our posture, but we might have to adjust our footing or shift our balance accordingly. We might even bend over laughing or need to sit down. If we are seated with our feet planted firmly on the floor, we may have the urge to stomp our feet as we laugh and our upper body is busy doing whatever it likes to do. If we are not seated properly, maybe we are leaning back in our chair or have our feet propped up, we might muscle contract our legs so much it thrusts us out of our chair. If we fall out of our chair, that puts us on the floor. And if we are on the floor continuing to experience limb muscle contractions, then yes, we can reach a point where we are literally rolling on the floor laughing. Now, try to do all that and still have the ability to type on a computer.

As a comedian there have been three times when I have made someone fall out of their chair laughing and I consider each a professional achievement. In fact, they are going on my next resume as some of my proudest accomplishments. The last time was just a couple of months ago, in Gainesville, Florida.

Gainesville is a beautiful central Florida college town that I have really grown fond of and every time I perform there I seem to be really well received. They don't have a comedy club, but there are weekly comedy shows at some of the venues. I was performing in a bar called Rockeys, normally known for dueling pianos. My set was going very well and I was having an awesome time just playing with the audience, and there was this guy laughing hard in the front row with his feet propped up on the stage. I commented and turned to address the rest of the crowd and just as I was about to make another joke, he leaned back a little too far and loudly planted himself firmly on the floor. Now he's on the floor laughing, and now I start making jokes about the fact that he was on the floor laughing. Suddenly that became the show for a few minutes and the audience, him included, just loved it.

Muscles don't just contract, they release so that they can contract again. When we are laughing our muscles contract and release, contract and release, repeatedly, so we don't sit there with our hands clenched in front of our chest looking like goof-balls after a single clap. And our arms, legs, and back are not the only muscles doing this, ever heard of a muscle called the sphincter? More specifically, I am talking about the urethral sphincter. It's a muscle that most of us would prefer kept itself nice and tightly contracted, at least in public. Now, I don't want to get gross, but you may already know that for some people, laughter can result in involuntary urination. As the phrase goes, some people can laugh so hard that the tears run down their legs. That's also not our only sphincter muscle. I've heard people say things like, "I was laughing so hard I nearly shit my pants!" Definitely, our slang reflects an understanding of this possibility.

I don't think I've made an audience member laugh so hard that they pee their pants . . . but I can't say for sure because that's not the sort of thing someone would admit to me during a show. I feel like this is the next milestone I need to achieve, the next step of success. So please, if you ever see me perform live and I do make you piss yourself, please tell me okay? Don't worry, I can keep a secret.

Another point I would like to make about laughter is that it is inherently social. Whenever I discuss this, I always think about the famous quote from poet Ella Wilcox, "Laugh, and the world laughs with you." There was a second half to that line, but it lies outside of my selective memory. Laughter is a social behavior and we are more likely to engage in it when others are around. In fact, neuroscientist Robert Provine has shown that we are thirty times more likely to laugh in a social situation than when we are by ourselves.[50] Just like sex, humor is better with other people around. The simple presence of other people increases our likelihood of laughing by a factor of thirty! I think that is pretty important to recognize because it is not just our likelihood of laughing that is increased with other people, but our sense of humor and perception of what's funny. We need other people around for both of those.

As an example, here is a scenario I think most of us can relate to. Try to remember a time when there was a movie in theaters, a comedy, that everyone was talking about but for some reason you didn't get a chance to see it or decided it would make a good rental instead. For me, this was *The Hangover*, the original, not those awful sequels (everyone is free to their opinion, but oh man, those were unwatchable). When *The Hangover* came out in 2009, it seemed like everyone I knew was talking about how funny it was. From the preview, it seemed like a movie I'd watch but one that I would wait for until it was available on DVD. Besides, I was busy trying to be a comedian performing at San Francisco open mics every night. Bottom line, I missed it in theaters and watched it in my home a few months later. I remember watching it thinking to myself, *I don't really see what all the hype was about.* Sure, there were some good parts, but overall I didn't think it was as funny as everyone was making it out to be. Now to be fair, I have since seen the light and today list *The Hangover* as one of my all-time favorite movies. The sequels are still awful.

Have you had an experience like that? Why was our initial perception of the movie so much different than everyone else's? I believe it is because we consumed it wrong. Movies are media designed to be consumed in the presence of other people: at a movie theater, duh. When we see it in a crowded theater, the laughter of the rest of the audience cues our laughter response and we have a much more enjoyable experience overall. When we watch that same movie at home by ourselves, on TV or even worse on our laptop or phone, we are denying ourselves the same experience. We are consuming it wrong, and in the case of a comedy we would probably be better off not watching it (if it is not going to make us laugh, what is the point?). If we are thirty times more likely to laugh in a social setting, *then we are thirty times less likely to laugh when we are alone.*

There are comedies that are designed for individual or small audience consumption; we have a name for these. We call them "TV shows." Most, certainly not all, comedy shows use some method to trick your brain into thinking there are other people around when you watch them. A lot are filmed in front of a live studio audience, so the audience's laughter is recorded along with the show. Some use laugh tracks to give the sense that there is an audience present. But either way, the effect is to give your brain the illusion that you are not sitting alone on the couch eating Cheetos in your underwear.

As more and more of us are consuming media at home, I suspect that sometime in the future studios will add laugh tracks to post-theatrical releases of funny movies. It makes sense to me, as long as the audience doesn't perceive it as a "Greedo shoots first" kind of change and it comes as an option for those who would prefer to turn it off. There, I just gave Hollywood a million-dollar idea. Now, I wait by my phone. In my underwear, eating Cheetos.

The presence of other people is definitely important to our perception of humor and potential for laughter, but the proximity of those people is also important. I learned this firsthand by conducting a series of informal "experiments" in my early days in comedy. As I shared in the introduction, comedy has been a lifelong dream of mine, but one that I postponed until later in life. There is no standard career path for comedians, but most of us begin by hitting the local open mics. If you've never been to one, a typical comedy open mic is held at a bar or at a comedy club on an off night. They give the comedians short sets, about three to four minutes, and generally try to squeeze in a lot of comedians, sometimes way too many (comedians tend to buy a lot of drinks, so it's good business). Because most of the comedians are just starting out, a lot of open mics are not very good entertainment unless you like watching dreams get crushed to death and guys talking about their penises. Open mics are generally attended by other hopeful comedians, who as a whole make very bad audience members for each other as they are so busy reviewing their own three minutes of jokes that they don't pay attention to one another. I am trying to paint a picture for you of an open mic scene that is not very audience friendly and not very nurturing to would-be comedians, and yet all of us begin there and eventually claw our way out of it to various degrees of success.

When I started performing stand-up, I knew that the open mic scene would be a very slow route to growth. Telling jokes in front of disinterested audiences for three minutes a night was going to take forever to develop and I was already a late bloomer. Plus, I figured nobody would ever book me as often as I would book myself, so I decided instead to start my own comedy show. I had a great opportunity to do so at a restaurant in San Francisco, and sure enough, that comedy show booked me a lot, like every night, even when I sucked. Having my own venue meant I could host the show and get as much onstage time as I wanted, in front of a real audience. It also taught me a few lessons about the comedy club business and the nature of laughter.

Initially we allowed people to just come in and seat themselves. For some reason, most comedy fans do not want to sit front and center, and most would avoid sitting in the front row altogether. Also, given an empty room, people tend to utilize the space, so they will sit with some distance between themselves and other people. Usually, the gaps would eventually get filled in and we would have a room with a decent-sized, packed audience. However, comedy is a tough business and there were many nights when the room was hard to fill. I noticed that on nights with small audiences, where the people sat influenced their reaction to the show. If we only had ten people in the room and those people were scattered throughout it, the show generated much less laughter than a packed house. Not only did the audience laugh less, but the show suffered because the comedians were not getting the feedback to keep them going. Then, we started seating people as they came in, filling the front rows tightly and expanding outward as people arrived. Now if we only had ten people, those people were seated in close proximity to one another. I noticed a huge change in audience response on those nights. The audiences may have been small, but they were mighty. They laughed more, the comedians received more encouragement, and everyone enjoyed the show more as a result of sitting in close proximity. Comedy clubs know this, their tables are usually small, and their chairs are crammed together. They seat their audience members tightly, and ask smaller parties to share tables. Through my informal "experiments" on seating, I basically learned what the rest of the industry had learned a long time ago.[51] Being in close proximity to other people encourages laughter. Laughter is contagious, and way better than syphilis.

The presence of other people increases our likelihood of laughing. This is why, I believe, that most uses of the acronym "LOL" are not accompanied by actual laughter. A lot of the time when I use that phrase, I am working on my computer at my desk with few other people around. My friend William just shared with me that the only time he laughs out loud when he types LOL is when he imagines sharing the joke with others. In other words, when he is imagining being social. Sometimes when I use the term, I may be in a public place, like a coffee shop, but the people around me are not exposed to the same stimulus and offer no incentive to laugh. For example, as I write this, William and I are sitting in a coffee shop on Fairfax Avenue in Los Angeles. There are a few other laptop jockeys here beside us and I bet if I burst into spontaneous laughter in response to something I saw on Facebook, I would probably be laughing alone. William might ask me what I find so funny, but the moment would be over. On the other hand, if something happens in the store that results in a shared experience, I bet most of the people in here would laugh. Well, all except that one guy, he looks a bit too into his screenplay.

I like to describe laughter as a product of its bodily components. Remember, I said that laughter is a condition in which we are simultaneously experiencing uncontrollable muscle spasms and desperately gasping for air. That doesn't sound very healthy, does it? And being in close proximity to other people tends to bring it out in us! Described like that, it starts to sound a lot like an orgasm, and most of us seem to enjoy those.

We don't all laugh at the same things, we don't laugh all the time, and we definitely don't roll on the floor every time we laugh, but laughter is incredibly good for us. The singer Reba McEntire once said, "To thrive in life you need three bones: A wishbone. A backbone. And a funny bone." I tend to agree with her. Although I do think she left out a fourth, very important bone.

LAUGHTER: WHAT'S THE POINT?

IN MY INTRODUCTION I MENTIONED THAT LAUGHTER HAS SOME real and important benefits. I am going to go into a detailed discussion of that material in the next few chapters, but the first and perhaps most important thing that I would like to discuss is how laughter influences our emotions, specifically our positive emotions. Smiling and laughing are not just responses to humor, they are behaviors we exhibit when we are feeling happy. Laughter makes us happy, and when we are happy we laugh.

That statement sounds simple enough, but consider the impact of social settings on laughter: we are more likely to laugh when other people are around. Now, if we take that a step further, we can infer that we are more likely to be happy when other people are around. I'm very fond of stating that outside of our own head, our greatest potential for happiness comes from other people. Obviously, not everyone in the world is going to make us happy; some people are just going to make us miserable, but the presence of other people in our lives increases our potential for happiness.

Have you ever suffered from depression or known someone with depression? Among other symptoms, people with depression experience less happiness in their lives as a result of the disorder. In my introduction I mentioned that the behavioral choices we make throughout our lives contribute to the development of the disorder. One of the behaviors that contributes to developing depression is social withdrawal, depriving ourselves of important social contact. This has a lot of influence on the development of the disorder, but one of the ways it impacts our emotional state is by reducing our exposure to other people. By doing so, depression reduces our likelihood of sharing a laugh with other people. It decreases the enjoyment we may otherwise experience by engaging in certain activities. When people have depression they often withdraw into social isolation as a function of the disorder. Therefore depression decreases the opportunities that depressed people have for laughter even further. When you have depression, you often don't feel like laughing and that makes sense, you are depressed.

Laughter is not a cure for depression, but it does elevate our mood. It is important to be aware of the impact of social behavior on our likelihood to laugh. Many people with the disorder have a real desire to recover. Some people with depression will expose themselves to comedic material to help counter their negative affect. This has been called the "humor therapy" movement and, as I previously mentioned, it is nothing new. In the sixteenth century, priest and religious reformer Martin Luther utilized a form of humor therapy when counseling people with depression. He advised against social isolation and encouraged them to surround themselves with friends that would make them laugh.[52] Today, we have a plethora of ways in which we can expose ourselves to humor while in the comfort of our home. Modern humor therapy often relies on TV, CDs, DVDs, or the Internet as an easy means to generate laughs—however we should not forget the importance of consuming these materials in a social environment. Watching a movie on our couch, by ourselves, is going to be far less impactful than watching that same movie with friends or family. One movie or shared experience is not going to suddenly snap us out of it. We didn't get depressed overnight and we certainly aren't going to be cured by a viewing of *Hot Tub Time Machine*—but it can help. And if we are working to recover, engaging in humor therapy, it is important to remember the value of other people. A wise man once said that outside of our own head, other people represent the greatest potential for influencing our happiness. If laughter makes us happy and other people make us laugh, then other people have a lot of potential to contribute to our happiness.

THE BENEFITS OF LAUGHTER:
WHY WE SHOULD CARE

W HAT WOULD BE YOUR REACTION IF I WERE TO TELL YOU THAT having a sense of humor and enjoying a good laugh on a regular basis could cure a variety of mental and physical illnesses? For example, mental illnesses like emotional disorders, personality disorders, addiction, and behavioral disorders can be eliminated through laughter. Sounds pretty good, eh? What would be your reaction if I said that laughter can also cure chronic pain, fibromyalgia, arthritis, diabetes, heart disease, obesity, and cancer? Those are some pretty serious illnesses and again, what if I claimed that all can be effectively cured by simply laughing? Amazing, I know! Would you believe that having a good sense of humor practically eliminates all suffering from flatulence or halitosis? Laughter has the potential to eliminate poverty and end world hunger. Laughter will teach you Spanish and even do your taxes for you. What would you say to that? It sounds good but I'd hope you would call me out on my bullshit. Clearly every one of those statements is either a gross exaggeration or completely untrue. Well, except for the one about taxes, laughter has an amazing grasp on tax law and a keen attention to detail.

I wish I could tell you that laughter was a cure for something, anything, but it is not. At best, it is a cure for silence, it may be a cure for the blues, but it isn't a long-term solution for anything. Self-help books offering quick and easy solutions to problems are sold all the time. Exaggerated claims are common, science is simplified for the masses, and we all like the promise of a quick fix. In one of my seminars, an audience member once asked me how to go about losing weight and I gave a very simple answer, "Decrease your calorie consumption and increase your physical activity." In other words: eat less and exercise more. She said to me, "Yes, but what ELSE can I do? I hate exercising and I love food." With the exception of surgical interventions (which still just force your body into a diet) there is nothing else to do. Despite the fact that obesity is epidemic in the United States, we all know exactly how to lose weight. There is no mystery there, eat less and exercise more, but we don't want to hear that. I just did a quick search on Amazon and found 22,700 books under the category "Diets & Weight Loss." That is a whole lot of ways to say something every one of us already knows. My point being that many of us are tempted to promote our ideas as the next big thing, offering a potential quick fix solution to consumers. And it apparently does sell books.

(By the way, before I lose some of you due to the above analogy regarding weight loss, you should understand that saying that the answer is simple is not the same thing as saying the execution of that answer is simple. It is actually extremely difficult to modify our behavior and the vast majority of us struggle to do so. I know how to lose weight, but as I've shared with you I am still carrying too much of it. Knowledge and action are two different things.)

I wish I could tell you that laughter would solve so many of our problems, but it won't and it doesn't. What it does do is make life better, and contribute to overall improvements in some of the illnesses I mentioned. It won't actually do your taxes. But it CAN help improve our emotional condition, it can help ease our pain, reduce our stress, and help prevent certain forms of disease. The operative term is "help." Laughter isn't a cure for anything, but it sure can help a lot. In the next series of chapters I will discuss some of the major benefits of laughter.

LAUGHTER MAKES US HAPPY

LAUGHTER MAKES US HAPPY. I SHOULD MENTION THAT NOBODY IS happy all the time, and we wouldn't expect that. In fact, to be happy all the time is indicative of a disorder.[53] That ain't healthy at all. However, there are people who are generally more happy than others, that's a given. And for a lot of us, laughter helps get us there. As Jimmy Buffett sang in his song "Changes in Latitudes, Changes in Attitudes," "If we couldn't laugh, we would all go insane."

This is my personal favorite benefit of laughter, the mood elevation I experience whenever I engage in it. I love to laugh and I laugh often. I'm also very happy; this could be both a function of how frequently I laugh and a reason for it. I am very fond of saying that laughter is simultaneously both an expression of our happiness and a contributor to our happiness. We laugh when we are happy, but that laughter also makes us happy. The second part of that statement is explained by a very old and robust psychological theory, the James–Lange theory.

You may recall that in my introduction I used a quote from William James, the father of American psychology. I will re-use it here, for the benefit of all you introduction-skipping speed readers: "We don't laugh because we're happy, we're happy because we laugh." He is also credited with a similar quote, "I don't sing because I'm happy; I'm happy because I sing." And his less popular, "I don't stink because I wear deodorant; I wear deodorant because I stink." Okay, that last one I made up but you have to admit it's quite clever.

In the two legitimate James quotes I shared, he is referring to his theory that over time has been referred to as the James–Lange theory of emotion. Despite the name, William James and Carl Lange developed their theories independently. History has chosen to credit them both although James is arguably the more quotable of the two. According to this theory, which I discussed in the introduction but I know a lot of you Quicksilvered it (yes, I did turn an X-Man into a verb) so here it is again, our emotions are experienced as a subjective interpretation of physiological feedback.

Whenever we encounter a stimulus, internal or external, our unconscious assessment of it may trigger a set of physiological changes in us and that physiological response in turn is interpreted by our brain as an emotion. As an example, suppose we see a big, growling wolf. One of the first things our brain has to do is determine if that stimulus is going to harm us in some way (and yes, it probably will because we all know what happened to the three little pigs). As I mentioned in my discussion of the brain, one of the areas important to carrying out this step is called the amygdala. I don't want to repeat myself too much, but if the amygdala determines that a stimulus is threatening, it sends signals to other parts of our brain that activate our stress response. The sight of that clearly threatening wolf triggers certain physiological changes to occur; among them, our hair will stand on end, our heart rate will increase, and our senses will become more focused. These physiological changes provide feedback to the brain, and our brain integrates this physiological information along with the sensory information of the sight of the wolf and interprets the experience as the emotion of fear. The physiological changes that take place ready the body for action; the emotional experience further increases the likelihood of engaging in an appropriate lifesaving or threat avoiding behavior.

Suppose we encounter a stimulus and our amygdala does not determine that it is a potential threat. Then it has another question to answer: Is this something that we may enjoy or might make our life better? If it is, then different signals are sent to other areas in our brain and we engage in a host of other physiological and behavior changes as well. These responses still provide our brain with feedback, and we interpret the experience as a positive emotion, such as happiness or love.

William James was unaware of the neurological mechanisms underlying his theory and it's not his fault—it was the nineteenth century and nobody understood them. Hell, when he was studying, people were still measuring the bumps on the head to understand personality (which does not work, in case you were wondering). One of the things I love about the James–Lange theory is that even while lacking our contemporary knowledge, he totally nailed it. Totes. In the past one hundred years or so we have learned a lot about the brain and confirmed that yes, in general, James was spot on. There are some variations on the theory to account for special circumstances, but in general our emotions are a product of our brain's interpretation of our physiology. This is how we feel our emotions.

Smiling and laughter are behaviors that are normally elicited by positive stimuli. Someone smiles at us, for example, and it prompts us to smile back automatically. Because our emotional experience is influenced by our physiology, that smile will then make us happier. I previously mentioned that there are a lot of studies that show we can elevate our mood by tricking our brain into thinking it is smiling. Again, my favorite demonstration of this is the one used by Strack, Martin, and Stepper (1988) who asked participants to hold a pen in their mouth either straight out like a cigarette or sideways in their teeth. Holding the pen straight out activates the facial muscles used in frowning, but the sideways position of the pen looks a lot like a smile, and the brain is tricked into thinking that it is suddenly smiling. Because smiling is a behavior associated with positive emotion, we start to feel more positive as a result of our fake smile and even respond to humor more.

This is my favorite way to demonstrate the James–Lange theory with a large group of people, and I would like you to participate as well, right now. First, take out a writing utensil or anything similarly shaped. Chopsticks work great, and in a pinch I've used my own index finger. If you don't have anything in your immediate vicinity, it is okay, I'll wait. Go on, get a pen! Got one? Good. Now, I want you to hold it using only your lips so that it sticks out like a cigarette. Hold it there just for a few seconds and then take it out, please. I don't want to make you miserable. Wait a few seconds and now take that same pen and hold it in your teeth sideways. I'm not positive this will translate perfectly in this format, given you have prior knowledge of why I am asking you to do it, but if you do it right, you might be experiencing a sudden elevation in your mood. In fact, go ahead and keep it in there as long as you want. Look at yourself in the mirror, or take a selfie while you are at it (and if you post it, be sure to tag @drbrianking). Even under normal circumstances, holding a pen in your mouth isn't going to make you jump out of your seat with joy, but you probably do feel slightly better as a result of tricking your brain into thinking it has encountered a smile-inducing stimulus. If done correctly, this minor manipulation of our facial muscles works. It's not just my favorite demonstration for James–Lange, it's also my favorite intervention for negative affect. If you are feeling down one day and you need a quick pick-me-up, stick a damn pen in your mouth!

Of course I know most of you reading this probably did not participate in the pen holding exercise. If you are one of these people, I'd like you to ask yourself why. Why didn't you do something that just a few paragraphs before I explained to you would have made you happier? Maybe you didn't have an appropriately shaped object to put in your mouth. Fair enough, but I am pretty sure you at least have a finger. Maybe you had a pen, but the thought of putting it in your mouth filled your head with anxiety about germs. Sure, germs exist, and they are no doubt partying it up all over the surface of your Bic disposable, but the chances of any real contamination are ridiculously low. Irrational fears, and the beliefs they stem from, can prevent us from engaging in all sorts of otherwise positive behaviors. Maybe you didn't do it because you are reading a book on your own as opposed to being in the presence of a room full of people also engaging in the exercise. Social context can have a lot of influence on our behavior, I get that. Whatever your reason, I'd like you to reflect on the fact that you are currently reading a section of a book all about how laughter makes you happy and for some stupid reason refused to engage in a minor behavior that psychologists have shown would have made you happy. Do I need to define irony?

When I do the pen demonstration with large audiences, an interesting thing usually happens. They will hold the pen in their mouth in the first position, cigarette-style, and despite the fact that everyone looks absolutely ridiculous, nobody laughs. Then, when they hold it in the smile position, we'll hear a faint wave of laughter throughout the room. It's usually no more than a giggle but it is worth noting that it never happens when the audience is holding their pens in the frown position and only occurs during the smile part. Smiling makes us happy, and when we are happy we smile and laugh. With this exercise, I observe daily a large group of people tricking their brains into thinking they are smiling, their brains then assuming that they are feeling positive, and that positive affect triggering a giggle response. As I am watching the crowd I will see eyes light up, reflecting the perception of the mood elevation. Again, it doesn't dramatically increase our affect, but it is noticeable. It's an awesome moment to share with people.

The pen "trick," as I call it in my seminars, really resonates with people. I get emails all the time from people who have attended my seminars and tell me how they share this with friends or apply this to their lives. For example, I once received an email from a woman in Alabama. She breeds and raises horses and told me about a night when she was on her way to a rodeo or some competition. She was pulling her trailer of horses along an empty, rural road in Alabama when she had some trouble with her truck. She pulled over and called for a tow truck. It took a while for the tow truck driver to find her, and when he did he realized his truck was not equipped to handle towing her truck plus the trailer of horses, so they had to call another tow truck. That truck ended up having a flat tire en route. There were some other details to the story, but the main point is that she ended up spending a disastrous night on the side of an Alabama highway and eventually missed her event. At the end of her email she said, ". . . and I went through five pens!" As much as that made me laugh, don't chew them.

By the way, I refer to this demonstration as a "trick" because it is exactly that. It works by tricking your brain into thinking it is feeling good. Have you ever been sad in the presence of another person and had that person say something like, "Cheer up, man, give me a smile." If you have, you'll know that smiling on command doesn't make us feel happier. When we voluntarily create one of those smiles, our brain knows it is a fake smile. It knows it is not associated with positive affect. How does it know this? Because it is producing that fake smile! The signal to move your facial muscles into a smile position is being generated by your prefrontal cortex and it is well aware that you are engaging in an insincere emotional expression. But when you put the pen in your mouth, you catch your brain off guard. It isn't expecting to be smiling at that moment, and it might take a moment to realize that it isn't smiling, but at least you had that brief moment of an elevated mood.

Like smiling, laughter and applause are other behavioral responses that are elicited by positive stimuli, and in turn make us happier. Another demonstration of James–Lange that I like to share pertains to the world of stand-up comedy, or the performing arts in general. As a comedian I am guessing that I probably go to more comedy shows than you, but if you have ever been to a live performance you might be aware of this. There is a trick that performers sometimes use to elevate your affect before the show. They come on stage and before they start telling jokes, they say something like: "Thank you so much for coming out and supporting live stand-up comedy. Give yourselves a round of applause just for coming out tonight!"

They ask us to applaud, and being polite audience members in a social setting most of us will oblige. Now the reason why this practice is so common is because audiences typically come in "cold." They are nervous, adjusting to their environment, getting settled in, etc. . . . and need to be "warmed up" or primed for the show. The act of applause, just like smiling, is something our brain associates with feeling good. We do it when we feel good, so if we are doing it our brain assumes that we feel good. It doesn't matter that we've been asked to applaud and it doesn't matter that technically we are applauding ourselves (what does that even mean, really?) what matters is that we are applauding and our brain receives that feedback and we start to actually feel good (or better). With an elevated mood, we become much more receptive to the show we are about to see. A lot of comedians do this, although I don't do it because I am genuinely funny (I am kidding, I so do this), and it is a common practice at practically every comedy club. Some may think of this as pandering and it can get a little tiresome if overused, but I will never fault a performer for using any tool at their disposal to better the mood of their audience.

I don't have any data to support this, but I suspect that comedians use this trick more often than other types of performers. Comedy is an art form that is specifically linked to positive emotion; we want to make you laugh so that you will feel good—and when you feel good, you are more likely to laugh. Other performers may wish to inspire positive affect through their work, but they are not limited to that emotion in the same way that comedians are. A dramatic play will not benefit from asking you to give yourselves a round of applause, neither would a dance recital or rock concert. However in comedy this is a common practice. Most comedians don't know that by doing this they are capitalizing on the James–Lange theory, but then again most comedians are not also psychologists.

When I share insight about this trick with my seminar audiences I sometimes worry that perhaps I've tainted them for their next comedian, and I definitely hope I have not done that here. If you go to a live comedy show and a comedian asks you to give yourself some applause, please don't defiantly sit there thinking, *No way, Mr. Funnyman, I know what you are trying to do.* The only person missing out in that scenario is you.

There is an entire industry that utilizes this practice to their advantage: television. Specifically, TV shows that are recorded in front of live studio audiences. As I write this, I have yet to appear on TV as a comedian, but I live in Hollywood and I love the entertainment industry. I've been a studio audience member on several occasions, and if you have never had the experience I definitely recommend it. It's very interesting to see firsthand how some of our favorite programs are produced. One of the behind-the-scenes details that interests me is that every show with a live audience hires a comedian to come out and warm up the audience before they begin filming. I love to watch this person work, as both a psychologist with an interest in humor and a comedian who would potentially really enjoy a gig like that.

I study these comedian warmer-uppers closely. It's my impression that these comedians rarely tell jokes. They may have a few quick ones they throw out at the audience, but instead of a thirty minute set of material, they do a whole lot of "give yourselves a round of applause." They are energetic, engage the audience, and repeatedly ask the group to applaud for various reasons. Everyone has their own approach to comedy, even warm-up comedy, but you might see something like,

"Where are you from? Pittsburgh? Wow, that's a hell of a commute; folks, give Pittsburgh a round of applause! How about you, sir, where are you guys from? Denver? That's the lazy man's way of joining the Mile High Club! Folks, give Denver a round of applause. Is there anyone here from Fresno? No? Good then we are safe! Just kidding, give Fresno a round of applause!" and so on . . . As this continues you can see the audience become more and more energetic, their affect increases, and by the time the star of the show finally comes out the audience is super excited and ready to be entertained. They heartily laugh at the jokes, they vigorously applaud the show, and even though there are flashing "applaud" signs in use, it all seems very sincere. They leave thinking they had the best time, smiling and talking much more than when they first arrived.

I have had experiences sitting in on shows where I thoroughly enjoyed myself, laughing my ass off in the moment, and then watched the same program later on TV and thought to myself. *Wow, that isn't nearly as funny as it seemed earlier.*

Most studio audience members are volunteers, but there are agencies in Los Angeles that recruit paid audience members for new shows or to fill seats when there aren't enough volunteers. Like any other wannabe in Hollywood, I am registered at a few such agencies. One is Central Casting, the largest casting agency for background actors or "extras" as they used to be called. This brings me to a really neat story regarding the effect of laughter on happiness.

If you are old enough, you may remember the sitcom *The Nanny* starring Fran Drescher (okay, you don't have to be that old, it aired in the mid 1990s). During its run, Fran became concerned about potential stalkers, and the producers decided that they no longer wanted to record in front of an audience of volunteers that was basically open to anyone. Instead, they wanted to recruit an audience of registered professional background actors who they felt they could trust. Lisette St. Claire, one of the casting directors at Central Casting, put out a call and decided to audition people based on the quality of their laughter. Eventually, she collected an audience of professional laughers and that became the regular studio audience for *The Nanny*. Over time, word got out about these exceptional laughers and they began getting booked to sit in on other shows. They were paid laughers and they absolutely loved their jobs. As you can imagine, laughing on a daily basis like that would generate a lot of positive emotion. Unfortunately with the onset of reality TV in the early 2000s, demand for professional laughers sharply declined and many of them reported increased feelings of depression as a result.[54] Laughter has been shown to elevate our affect, even when there is nothing to laugh at. Voluntary laughter, that is, laughter we engage in on purpose, provides the same benefit as laughter that is more organically inspired. There is an extremely popular international movement that uses this as its foundation; it's called "laughter yoga." It was started in 1995 in Mumbai, India when a physician named Madan Kataria recognized the physical and emotional benefits of laughter. He gathered together a group of five participants to form the very first "laughter club" and today there are more than eight thousand clubs spanning across sixty-five countries.[55] The practice includes exercises based on clapping, breathing, and of course laughing. The laughter starts off fake and feels very fake at first, but that fake laughter becomes real laughter fairly quickly. I have seen video of sessions starting out with breathing exercises and ending with loud uproarious laughter for an extended duration. Laughter yoga is an exercise. It is laughter without amusement, but practitioners report many of the same benefits that derive from laughter inspired by humor. I like to recommend laughter yoga, or similar practices, to anyone who has difficulty laughing organically. According to Dr. Kataria, "We are paying a very high price for taking life seriously. Now it's time to take laughter seriously."

Other practices have sprung up that are similar to laughter yoga, with names like "laughter meditation" and "laughercize," but the core principle is the same: all laughter, even voluntary laughter, is beneficial.

When we are happy, we smile and laugh. Smiling and laughter, two behavioral responses to humor, also make us happy. William James observed this over a century ago and scientists ever since have been confirming it in research. Knowing that this is the case, that our emotional condition is influenced by our physiology, gives us an opportunity to consciously modify our affective state through laughter. Hollywood has been using this basic phenomenon for decades and we can as well. Find something to laugh at, take life a little less seriously, and we'll be on our way to living a happier, healthy life. Laugh, laugh a lot. Even if you don't have anything to laugh at.

OKAY, SO WHAT'S SO IMPORTANT ABOUT HAPPINESS, ANYWAY?

I JUST SPENT A FEW PAGES DISCUSSING HOW LAUGHTER MAKES US happy, but to be fair nobody needs to be happy. Increasing positive affect does help alleviate depression, but clinical depression is not just about being unhappy. Despite the fact that depression is an increasingly popular disorder, it still affects just a minority of our population, meaning that most of us are not at risk of developing the disorder. Besides, there are plenty of people out there who live perfectly productive and healthy lives that are miserable as hell the entire time. And sometimes they live a little too long. Misery can fuel creativity and it can lead to innovation. It can be the motivation to create a better life. So if most of us are not depressed, and the experience of misery can occasionally have positive consequences, then is there any real reason to pursue happiness through laughter?

I think there is, and I am definitely not alone. In my introduction I mentioned that since 1990 the subfield of positive psychology has been growing within the discipline and influencing researchers in many other fields. Positive psychologists are concerned with quality of life issues, and specifically how we can improve it. Affect is a central topic for positive psychologists and we have learned a lot in the past twenty-five years thanks to our new scientific pursuit of happiness. We have a good understanding of what contributes to our happiness, and how happier people differ from less happy people.

For example, we know that happier people are more social and more energetic. Social behavior also makes us happy, and it is hard to feel much joy when you are lethargic, so it is important to understand that each of these benefits functions as a feedback cycle (i.e., happy people are more social and social people are happier). Happier people are also more resilient, that is, they recover faster from adverse events. When something negative happens to us it is natural to feel bad about it; not doing so is probably indicative of a serious psychological issue. But when a happier person experiences a tragedy, they will get over it sooner than a less happy person. Again, being resilient is also a contributor to their happiness.

Happier people are more likely to get married than less happy people.[56] This might not be that important for everyone, but it is worth noting that if marriage is a goal we increase our likelihood of achieving it by being happy. Very often I hear people say things like, "I'd be happy if I could just meet that special someone to share the rest of my life with, maybe off of Craigslist or Tinder . . ." Falling in love does increase our happiness, but if we require it to be happy we are probably more likely to make the other person miserable. Research shows that if you want to get married, get happy first. Smiling is attractive.

Happier people are also more likely to stay married once they get there. This makes sense as you find very few people who are happily married ending up divorced. You almost never hear, "You know, honey, I love you. This partnership we have entered has been mutually beneficial and very satisfying. You compensate for my many shortcomings and I compensate for yours, of which there are very few. We've grown together; we have raised some children together and have planned a great future together. But you know what, how about we call it quits?" No, you never hear that. Happy people don't get divorced; people get divorced TO BE HAPPY.

On a side note, I have never understood most people's reaction when someone gets divorced. People will often say they are sorry and offer condolences. I always say congratulations, because clearly that person is taking a step toward increasing their happiness. Sure divorce is a tough thing to go through, but it can be a necessary change.

Finally, happier people are healthier people. And remember, this also works in reverse so that healthier people are happier. As I am in the midst of writing this section, our current US surgeon general, Dr. Vivek Murthy, recently stated in a public address the value of happiness:

> Happiness affects us on a biological level. Happier people live longer . . . As we grapple with the challenges of how to create a healthier world, let's remember that happiness is a powerful tool to improving health.[57]

Dr. Murthy is correct, although he needs no confirmation from me of course; research shows that happier people are healthier and that they live longer than less happy people. Life expectancy relates to happiness. I find this very encouraging because I would like to live as long a life as I can (within reason considering some of my other less healthy habits). There is a lot of research that demonstrates this, but one of my favorite studies to cite is from Ernest Abel and Michael Kruger (2010).[58]

The study is sometimes referred to as the "Baseball Card Smile Study" because the researchers used baseball cards from the 1952 Major League season. They removed the photos from the cards so that they were separated from the players' names and statistics and presented the photos to a panel of modern participants. The participants then rated the quality of each player's smile. The reason for using baseball cards was that the photos were fairly uniform and information on each player is available in the public record. Looking only at data for players who had died prior to July 2009, they found that smile intensity correlated with how long that player lived. In fact, players with the most intense smiles lived on average seven years longer than players who did not smile. That is a seven-year gap in life expectancy predicted by a single photo taken sixty years ago. I think that is pretty amazing.

But being caught smiling once is not the same thing as being a happier person. It is possible that some of the intense smilers might have just had their contracts renewed that morning or found out they were not the father. It is also possible that those who did not smile were caught on a bad day (maybe they also found out they were not the father). However there is something very notable about happy people: we smile more often than less happy people. When we smile more often naturally and a photographer asks us to smile for a photo, we convey a subtle difference in the quality of that smile. That subtle difference relates to how long we are going to live. The fact that any gap in longevity is related to a single smile is pretty incredible. The bottom line is this: SMILE FOR ALL YOUR PHOTOS! You never know if that's the one they are going to pick! You don't want to shave seven years off your life because you were taking it too seriously one day, do you?

Positive psychology has shown us that there are a lot of factors that contribute to and increase our happiness. Our social behavior, including how often we express positive emotions like gratitude and love, our health, how well we manage stress, and our level of optimism are a few things that help make us happy. Those topics go beyond the focus of this book, but as I've discussed, humor is also a major contributor to happiness. Humor, and the laughter it inspires, also impacts the quality of our social relationships, and as we'll see in later chapters can make us more optimistic and alleviate stress. In other words, humor and laughter both directly and indirectly impact our happiness. And happiness is really good for us.

When I first set out to write this book, I had two goals. The first was to provide you, the reading audience, with some valuable and useful information. The second was to do so while being as humorous and entertaining as possible. From my perspective, both goals are equally important. The information is important, but I am not sure how well it might be applied. However, if I make you laugh at all while reading this book, then I consider it a job well done. By making you laugh I've slightly increased your health, and maybe just maybe added a little time to your life. You are welcome.

LAUGHTER FEELS GOOD

HAVE YOU EVER LAUGHED? OF COURSE YOU HAVE. EVEN IF IT has been a while, you already know that laughter feels good. Few people would argue that laughter is not a pleasurable activity. Okay, it may not be pleasurable to be laughed at, but that's something different. If we are the ones doing the laughing, we are experiencing pleasure. We could even argue that it feels good to be happy, and since we've already read that laughter makes us happy, laughter in turn makes us feel good.

Earlier I promised that I was not going to go into too much detail on the brain, and I meant it. However, in understanding how laughter is pleasurable, I need to discuss the brain a little more. Only slightly more, I promise. If this isn't for you, feel free to skip ahead to the next chapter, on how laughter decreases pain.

Laughter stimulates activity in areas of our brain that are associated with reward and pleasure. These are located in a set of subcortical structures called the limbic system, which also includes the amygdala. Subcortical obviously implies that these areas are located below the cortex, and as such are clearly not part of the prefrontal cortex. I previously described the prefrontal cortex as "the home of our conscious mind" (King, THIS BOOK) so you should be able to infer that the structures in the limbic system lie outside of our conscious awareness. We aren't aware of their activity, but that activity motivates and influences our behavior all the time.

I am not going to get too technical, but there is a part of your limbic system that many people commonly refer to as the "pleasure center." See? That doesn't sound technical at all. This is not the structure's official name of course, but the term "pleasure center" may refer to different areas, depending on who is using it and how detailed they wish to be. Regardless, the term always refers to at least the nucleus accumbens, a structure that is centrally located within the brain and receives inputs from lots of other areas including the previously mentioned amygdala. Activity in the nucleus accumbens increases whenever we experience something pleasurable. All drugs of abuse, from marijuana and alcohol to the hard stuff, have this affect. Certain types of food stimulate its activity, as does viewing pleasant images and listening to certain types of music. Sexual behavior, of course, stimulates activity in the nucleus accumbens and so does practically any experience we may consciously associate with a feeling of pleasure. Even exercise, so I've read, stimulates activity in this area. Unless you missed the title of this section, it should not surprise you that laughter also stimulates activity in the nucleus accumbens.

Note, recently neuroscientists have determined that activity in the nucleus accumbens serves more as a means to motivate our behavior toward a possible source of pleasure, and does not actually provide us with the pleasure we experience.[59] However, I think it is fine for our purposes here to still refer to it as the pleasure center. Whether its activity is providing us with the pleasure we experience, or simply signaling other areas and systems to provide it to us, pleasure is a big motivator.

Laughter increases activity in the so-called pleasure center of the brain; although this activity may not be the source of the pleasure we feel, it is definitely related to it. One additional effect of laughter on the brain is that it stimulates the release of chemical compounds called endorphins. Endorphins are a type of neurotransmitter, a chemical that is used by the brain to communicate between cells or groups of cells. Endorphins are also referred to as "endogenous opioids" because they are naturally occurring chemicals that have similar properties to opiates like morphine and its over-the-top buddy heroin. In case you've never been in serious pain or seen an R-rated movie, opiates (including the endogenous kind) have several effects on the brain and body. These include analgesia (many opioids are used as prescription painkillers) and euphoria. Not that I know from experience, but apparently opioid use can feel really, really, really good. This is one reason that opioids, those of the non-endogenous variety anyway, have a high potential to become habit-forming.

Did you catch that? Our brain naturally produces chemicals similar to morphine and laughter stimulates their release. You don't even have to know a guy. That's pretty awesome.

Now, the brain is an extremely complex organ and I am not going into nearly enough detail to do this justice, but here's an interesting tie-in to the pleasure center of the brain. I mentioned previously that substances commonly abused for their pleasurable effects all increase activity in the nucleus accumbens. Recent research has shown that the so-called pleasure center may not be the true source of the pleasure we experience. There are several areas in the brain that release endorphins, and perhaps the activity of the nucleus accumbens signals its release from these areas. Just like any other pleasurable stimulus, laughter may increase activity in the nucleus accumbens, which in turn may signal the release of endogenous opioids. Oh yeah, and to further confuse you, the nucleus accumbens totally releases its own endorphins, at least after exposure to alcohol or cocaine.[60] So, there's that.

You know, maybe we all should have skipped ahead.

DO WE REALLY NEED TO DISCUSS
THE IMPORTANCE OF PLEASURE?

I AM PRETTY SURE THAT I DON'T NEED TO BUILD AN ARGUMENT AS to why pleasure is a good thing. Pleasure is defined as a good thing. We like pleasure, that's part of the deal.

But, I will say that of all the things that we might find pleasurable, laughter is pretty great and less likely to ruin our lives if we over-indulge. Remember when I discussed the so-called pleasure center, I mentioned that activity in this area increases whenever we engage in behavior we find pleasurable. That includes drinking alcohol, using drugs, eating high-caloric foods, and engaging in sexual behavior. Well, some of the pleasurable experiences I just listed have the potential to become habit forming to the point that we may say that a person has developed an addiction. And by "some of" I really mean all. Laughter can be habit forming too, but very few people lose their jobs or outgrow their clothes because they laugh too much. Provided that our laughter is appropriate, laughter can be a healthy source of pleasure.

Also, pleasure makes us happy. And I seem to recall something about that being very good for us.

LAUGHTER DECREASES PAIN

I PREVIOUSLY DISCUSSED HOW HUMOR CAN BE AND IS USED AS A defense mechanism. That is to say that humor helps us deal with memories and thoughts that may be too difficult to express in a more straightforward manner. Sometimes the only way to effectively cope with a tragedy is to laugh it off. And why not? Tragedy generally makes us unhappy, laughing does the exact opposite of that so it seems natural that we would laugh away our pain. Sometimes there is nothing we can do but laugh at our misfortune.

I also discussed how we find benign violations to be funny. Some of those can involve somebody, not us, experiencing some real pain, real physical pain. Someone takes a pie to the face and we find that funny. Larry and Moe of the Three Stooges lift Curly off the ground by each grabbing onto one of his ears and yanking on them and we inspire generations of horrible mothers. A cartoon bunny shoots a cartoon duck in the face with his own rifle and we laugh our asses off as the duck's disembodied bill spits out lispy wisecracks. The guys from *Jackass* can Taser a dude in the testicles and make millions. You get the idea: we regularly laugh at pain, especially the pain of others.

But what if we are the ones in physical pain? Can laughter help relieve pain in this sense? Yes and no.

Before I go into this discussion I feel I should mention a man that we owe a great deal to for the research that follows, Norman Cousins. Norman was not a psychologist or a medical doctor but instead was trained as a journalist and had quite an amazing career. His story has been told by more knowledge-able folks than me, but let me give you the gist of it and explain why I bring him up.

Norman was on assignment in Moscow sometime during the mid-1960s, so pretty much smack in the peak of Soviet Russia. I'm trying to fit in a good Yakov-inspired Soviet reversal–style joke here, but it is just not coming to me. Anyway, while there he developed a condition that caused his joints to become rigid. He lost a great deal of his mobility and suffered from tremen-dous pain. Being in Moscow, he was taken to a Russian hos-pital where he was diagnosed and treated for a rare form of arthritis. Given the nature of his diagnosis, he was not expected to survive very long and he was unresponsive to the medica-tion he received. Also, the Moscow hospital environment was a stressful one with noise and people and constant activity. Okay, I think I have it now: In America, stress sends you to the hospi-tal. In Soviet Russia, hospitals stress you! Totally nailed it.

Being unresponsive to treatment in a highly stressful envi-ronment and practically dying anyway, Norman decided to check himself out of the hospital and into a nearby hotel. In the hotel, he could relax and he developed his own treatment plan based on some of his previous research. For years he had been reading about the influence of emotion on our physical health and believed that laughter and a positive attitude could help beat his illness. Also, for some reason, massive quantities of vitamin C. Like, so much vitamin C. This ain't a book about vitamin C, so we'll just deemphasize that little piece of the story.

He arranged to have a movie projector brought into his hotel room and got his hands on whatever humorous film clips he could find. These included Marx Brothers movies and epi-sodes of *Candid Camera*. I have always assumed his selection was influenced by limited availability in Moscow at the time and not necessarily his personal tastes. I could be wrong, and who knows, maybe a modern-day Norman Cousins would lie in his hotel room giggling while screening episodes of *Punk'd*. Anyway, based on his prior readings, he prescribed himself a daily dose of humor and made a point to laugh for at least ten minutes a day. He detailed his experiences treating himself in his book *Anatomy of an Illness,* and in it he wrote, "I made the joyous discovery that ten minutes of genuine belly laughter had an anesthetic effect and would give me at least two hours of pain-free sleep." That's a pretty incredible claim.

Ultimately Norman recovered quite a bit from his unexpected illness and went on to live for twenty-six more years, way longer than his Moscow doctors initially expected and way before Obamacare. He attributed his recovery to his self-prescribed laughter treatment, and claimed that "each patient carries his own doctor inside him." His book went on to become a made-for-TV movie starring Ed Asner, coincidentally a man originally made famous for playing a journalist.

So that's about the gist of why we usually discuss Norman Cousins, or at least why I do. Now that I have shared with you a pretty amazing story about how laughter can cure a pain-related illness, let me promptly shatter that illusion. In Russia, illusion shatters you. Okay, maybe that one doesn't work. Anyway, the main issue with Norman's claim comes from his diagnosis, in that we do not know what he had. When he arrived at the Moscow hospital he was given an initial diagnosis based on his presenting symptoms and began exploratory treatment. He was unresponsive to his medication, which could mean that the medication was failing him or the diagnosis was in serious need of revision. However, we will never know exactly what illness he was suffering from because he checked himself out of that hospital before his progress could be reevaluated. Without knowing for certain what he was suffering from, it is impossible to attribute his successful recovery to his self-prescribed treatment. We do not know if his recovery was due to laughing his ass off at the Marx Brothers or if it was just the natural progression of whatever he had. If I owned stock in the Marx Brothers catalog, I know where my opinions might lean in that discussion but unfortunately, we may never know.

The reason I like to discuss Norman Cousins is not because his case serves as proof that laughter can overcome pain, or that *Candid Camera* was an incredibly underrated program. His story may have its controversies, but I like to mention him because of what he inspired. In making such claims, he encouraged a generation of skeptical researchers to scientifically test the effects of humor exposure and laughter on our perception of pain. And it's a good thing they did too, because we now know, through empirical research, that a daily exposure to humor can reduce pain. There are many studies that have shown this effect. The effects are temporary, and laugher is not going to cure anything. I cautioned against this type of thinking earlier, but responding to humor can reduce pain and make life a little easier for those who suffer.

In the chapter "Laughter Feels Good" (literally it was just a few pages ago), I joked that some of you might want to skip a few paragraphs on the brain. Hopefully you recognized I was joking (remember I taught you how to do this much earlier in this book) and continued reading anyway. If you read that chapter completely, it should not be much of a surprise that laughter has analgesic effects. If you skipped then, surprise! Laughter has analgesic effects!

As I discussed, we have chemicals in our brain that bear more than a passing resemblance to morphine, the endorphins. In fact they are named for their similarities to the opioid painkiller; the name is a contraction of "endo-" for endogenous, meaning from within the body, and "-orphine" for morphine. Not to be confused with "Morphin'" which is something the mighty Power Rangers do on occasion (who am I kidding, they do it every episode). I suspect morphine's reputation as a painkiller precedes its appearance here and you are already familiar with it as a very effective analgesic. Your brain produces naturally analgesic substances similar to morphine, when you freaking laugh. I've said it before and I'll say it again: that's awesome.

However, not all of the research that has shown a reduction of pain sensitivity associated with humor and laughter ascribes the effect to endorphins. I will discuss these in a bit, so, you know . . . relax.

You may be wondering how we study pain sensitivity in experimental settings. The overwhelming majority of research in psychology is conducted on lab rats, or college students who are very similar to lab rats. They have the same basic diet. Actually, there is nothing special about being enrolled in college that makes someone a favorable research participant. Students are just commonly used because most researchers are university professors; they are a sample of convenience. Also, the fact that many of them could use some extra credit means that there is never a shortage of willing research volunteers hanging around the psych department. I mentioned earlier that rats laugh, but not in response to humor. Unless you happen to know some good rat jokes, that is. But you try getting them to pay attention. Besides, we are limited in what we can measure in rats. Therefore, for most researchers interested in studying the effects of laughter on pain sensitivity, human suckers, I mean participants, are the way to go. I mentioned that out of convenience most of these tend to be college students.

All research conducted and published in psychology (and any academic discipline for that matter) has to adhere to certain ethical standards regarding the treatment of our participants. When we study human beings, we have to be careful not to subject them to any procedure that might cause them permanent harm. This makes the study of pain a little challenging, because we kinda have to cause people pain in order to study their reaction. Therefore researchers have developed ways to induce pain in their participants without causing any tissue damage.

My favorite method is called the cold pressor test and it is a simple design. Basically, we take a bucket of ice water and ask our volunteer to submerge their hand up to the wrist. You can imagine that when they put their hand in a bucket of ice water, it is obviously going to be cold. Then, as they continue to hold it in there, that cold sensation will eventually become painful. At the precise moment the experience becomes painful, we ask them to let us know and record how much time has passed. That becomes our measure of their pain threshold, how many seconds until the sensation turns to pain. After recording their pain threshold, we ask them to keep their hand submerged for as long as they can stand it (or up to some maximum time limit, like five minutes). When they decide that they absolutely cannot stand the pain any longer, they withdraw their hand and we record the time. That is our measure of their pain tolerance, how many seconds from submersion to withdrawal. Both pain threshold and tolerance are measures of pain sensitivity, and we all vary in our sensitivity to pain. The nice thing about the cold pressor test is that it does inflict pain, but it is a temporary condition and the research participant leaves the experiment unharmed. We get to learn about pain sensitivity without getting all stabby.

It's also cheap: a bucket and a stopwatch. Even a grad student can afford that.

In my own experiments, conducted mainly in graduate school in New Orleans and Bowling Green, Ohio, I found that both men and women have an average pain threshold of about twenty seconds and can tolerate the ice water for an average of about sixty-seven seconds.[61] However, there are individuals we call outliers who can hold their hands in the water much longer and reach the five-minute maximum time limit. It has been my experience that more men than women reach this limit. One guy told me that, as an athlete, he regularly ices down as part of his training and had become desensitized. Others attributed their tolerance to surviving harsh winters; this was northwest Ohio where I once celebrated spring break with a snowstorm that froze Lake Erie. Worst beach day ever. I still got burnt. Wear that sunscreen, people.

In a recent series of experiments using a variation of the cold pressor test that substituted a frozen wine-cooling sleeve for a bucket of ice water, anthropologist Robin Dunbar and his associates found that laughter elevates pain threshold.[62] They tested participants watching videos in the laboratory and enjoying live performances at the Edinburgh Fringe Festival, which is a massive annual event in Scotland that includes a lot of comedy. The effect of laughter on pain sensitivity, which they attribute to endorphin release, was even more pronounced when participants were tested in groups. You might remember that earlier I mentioned that we are more likely to laugh in the presence of other people; well, those other people are also potentially influencing our endorphin release.

"The human race has only one really effective weapon and that is laughter," said Mark Twain. "Also, did you ever notice how much I look like Einstein? He and Colonel Sanders are totally stealing my style." Okay, I may have embellished that last part. I use that legit Mark Twain quote a lot, but as a weapon, laughter is more of a defense. Despite having many studies that show it can help reduce pain, this does not seem to be widespread knowledge, or at least published reports of applications of this weapon seem limited.

Most commonly, I find accounts of using laughter in the field of dentistry.[63] This makes a lot of sense, dentists' offices are basically torture chambers and many of the procedures we go in for can be quite painful. Humor and laughter are very often used by dentists and staff to help patients relax and put them at ease. The more relaxed a patient is, the less pain they are going to experience.

I have mentioned that my seminar tours are attended by a lot of health-care professionals, and this includes dentists, dental assistants, and hygienists. My sample is not at all representative, but I always find dentists to be very funny. I hear a lot of dentist jokes, most of them told to me by dentists themselves. Many hygienists who attend my seminars tell me they work for a funny dentist and that he or she is always cracking them up around the office and making their patients laugh. One dentist explained to me that he was a "forced extrovert." Normally introverted, he knows how important establishing a rapport with a patient can be, so when he enters the exam room he turns on the charm. He starts with a joke, gets a quick laugh, and continues to go out of his way to behave in a manner contrary to his outside behavior. He said that when he is with a patient, "it's like being on stage" and that he puts on an act. As a comedian, I can definitely relate to that.

Like I mentioned, dentists are funny. From all the academic journals I have reviewed, I have really only encountered something like this in a dental journal—a joke used to help reduce patients' pain:

A bout of stomach trouble had sent a man to the physician. The physician prescribed plenty of milk and gave the man a bottle of pills. "I'll stop by this evening and see how you're doing," the physician said. "In the meantime, drink at least four glasses of milk. Milk is the ticket for curing your trouble. So drink plenty of it." That evening, the physician returned, examined the man and told him, "You're much better this evening. Just be sure you don't drink any milk. Not one glass. It's not for you." "But, doctor," the man exclaimed, "only this morning you told me that milk was what I needed and that I should drink four glasses of it." "Well, what do you know?" the physician replied. "It certainly goes to show that we've made tremendous progress in medicine since the last time I saw you." (Morse, 2006)[64]

Yup, dentists are funny.

Comedic videos are sometimes used to help dental patients relax while they are being anesthetized. The calming effect is primarily thought to be a result of distraction. In fact, there is evidence that shows that even without an endorphin increase; laughter helps reduce our sensation of pain simply by distracting us from it.[65]

I cut my thumb off once. Well, who hasn't done that at least once, am I right? Actually I sliced off the tip of my left thumb one night. It was a stupid accident. Sometimes I think I am an artist and one of those times I was working on a plastic sculpture project that really consumed me. After working on it all day, I went to bed to get some rest and was so excited about the project that I just couldn't sleep. I got up in the middle of the night and started slicing away with my X-Acto knife. A little groggy and not paying much attention, one of those slices included the tip of said thumb. I remember the instant pretty well, as if life suddenly changed to slow motion. I could see the blade enter my flesh and slowly move through it in a clean slice. I watched as the tip of thumb, formally attached to my body, flew off the knife and into the air in a small arch toward the floor. I watched it hit, completely dumbfounded. Then my conscious mind caught up and I realized, *oh shit, I just cut off my thumb*! I reached down to pick it up, and holding a piece of my left hand with my right I stared at my now exposed thumb bone and reflected on how it didn't seem at all painful.

Don't let me fool you though; it got painful in a hurry. Sometimes it takes a moment for the pain to kick in and once it did it was excruciating. I am a big baby after all. With my piece of thumb now in a plastic bag on an ice cube, my roommates gave me a ride to the emergency room. I sat in the back of the car and I remember the entire time repeating the words "I cut off my thumb" over and over as if in shock over the whole situation. To drown out the sounds of my pain, the driver turned on the stereo and popped in a comedy CD. (He was a comedian too, and had a few he picked up recently.) A few jokes in and I started to pay attention. I stopped repeating myself and even laughed a bit on our way to the hospital. It still hurt like a bitch, but the distraction helped relieve the pain. I'm sure my roommates were equally relieved that I shut up.

IS NO PAIN OUR GAIN?

LAUGHTER MAY STIMULATE THE RELEASE OF NATURALLY PAIN-killing endorphins. In addition, humor gives us something else to focus our attention on and in doing so can help us calm down. If you have ever done so much as stubbed your pinky toe on the leg of a coffee table, you already have a good understanding of the value of pain reduction.

Norman Cousins claimed that his form of humor therapy helped him recover from a potentially terminal and painful illness. As I described earlier, we don't know what specific illness he was suffering from so we can't really evaluate the validity of his claims of a cure. However, I am certain that he did experience momentary relief from his pain after each one of his laughter sessions. Modern research definitely supports that likelihood. Laughter might not have cured him, but it definitely offered an effective form of pain management.

Once again I would like to stress that laughter is not a cure for anything, but it can help make life better. It can be a valuable tool in fighting a condition we refer to as "chronic pain syndrome." There are many forms of chronic pain and I imagine that none of them are very fun to have, but generally speaking chronic pain syndrome refers to pain that should have gone away but hasn't. Let's say a patient suffers some injury, and that injury is the source of significant pain. Well, with the natural healing process the pain should eventually go away, but for some reason it lingers much longer than would be expected. The pain might also be more intense than we would expect. Obviously not all forms of chronic pain are expected to heal, therefore distinguishing chronic pain from chronic pain syndrome. I know it sounds confusing, but you'll catch up. I have faith in you.

So what's the difference? Well, chronic pain syndrome appears to be a behavioral disorder. That is, the patient's affective and behavioral choices influence their experience of pain. Pain sucks. Nobody likes being in pain. Okay, masochists like being in pain, but that is something entirely different, and a different kind of pain. Generally speaking, pain is not an enjoyable experience and it affects our mood. Pain can make us miserable, and reasonably so as it is hard for most people to remain cheerful and positive when their body is full of sharp stinging sensations. However, we do not all react to pain the same way. Some people allow it to seriously influence them and they become negative, mopey, depressing throwers of pity parties (and trust me, those are the worst parties). Their behavior looks a lot like that of people suffering from depression, and that depression they experience further agitates their painful state by reducing their joy and interfering with recovery. You may have heard that a positive, optimistic mood has been shown to aid in recovery; well a negative, pessimistic, nasty mood can do the opposite. That's when the pain becomes a syndrome.

Our behavior has a huge impact on our mental and physical condition. As an example, let me tell you about one of my uncles. I mentioned previously that I am originally from New York. Although I am currently in Los Angeles, most of my family still lives back east. When I was growing up I had this one uncle who moved away from the rest of the family and settled in California. At the time, I had never been out there and my young mind was filled with images of movie stars, palm trees, and bikini girls on beaches. He was my cool uncle in California.

Later, when I did eventually move out west, I visited the town he had lived in, a place called Fresno, and my uncle became a lot less cool in my eyes (although to Fresno's credit, it at least has palm trees). Nonetheless he was out here living the dream. He had his own business and bought a home, and every time I would see him in New York he seemed genuinely happy.

At some point though he started experiencing bouts of intense pain in his lower face. He was diagnosed with a condition called trigeminal neuralgia (TN), which affects the nerve that provides sensation for most of the lower face. The pain associated with this condition has been described as just awful and has led to countless uses of the old joke, "Does your face hurt? Because it's killing me!" by smartass younger brothers everywhere. It is a difficult condition to treat, and my uncle attempted many such treatments with no relief. Eventually he decided being in pain so far away from his family was less than ideal and he sold his business, and picked up and moved back to New York. He settled in the attic apartment of his sister's house and that is where he's lived for the last few decades.

When he first moved back, I think he was happy to be around family again, and he was active in their lives, socializing with my cousins and laughing. It is probably not a surprise that my family is big on the laughter. However, still experiencing the pain, some days he found it hard to motivate himself to join the others for meals and he started spending more and more time alone. Over the years, his behavior gradually became more and more withdrawn, and he became very socially isolated. He did not seem to have any friends and would barely communicate with other family members. Whenever he did speak, he was nasty and negative. This eventually led to others avoiding him, and even some serious animosity among family members. My cool uncle in California had become a miserable attic-dwelling jerk.

Now I was not around to witness this transformation; I was out in the rest of the country getting my education and working on my career. But every once in a while, I would make it out to that house on Long Island and catch up with the family. I would almost never see my uncle; each time he would stay upstairs my entire visit. That is, except for my last visit.

I was in the living room sitting on the couch relaxing and watching a little TV when my uncle surfaced from his man cave up the stairs. Given all that I had heard from the years before I was shocked when he started up a conversation with me by saying, "Hey, nephew." (Appropriate sure, but what a weird way to address me.) He sat down on the couch next to me and we started talking. I told him about the direction my career had taken me, and how I had been traveling the country speaking about topics pertaining to happiness, all of which I gathered he had already heard from others. Then he opened up.

"I think I am depressed," he said. "It's this pain, dealing with it has made me miserable and I am not sure what to do about it."

I responded with "well, duh!" and before you think me heartless remember this is my uncle, not a patient. Bedside manner is a bit different for family. And so is the co-pay, he still owes me.

We talked about it and he told me about how he has tried this and that treatment, surgery, medication, etc . . . and nothing seemed to provide him with relief. He blamed his depression and mood shifts on his pain and wanted to know if I had any insight. Without going into too much detail, I explained to him just as I explained a few paragraphs ago, that our behavioral choices can influence our physical condition and even prolong or amplify the sensation of pain.

I said to him, "It sucks that you are in pain. I understand that, and you didn't choose to be in pain. However, you are choosing how you cope with that pain." With social isolation not only was he missing out on interactions and activity, but he was also missing out on joy, laughter, and all the benefits associated with those experiences. I went on and explained, "You are going to be in pain whether you are lying in bed, alone in your sister's attic or out in the world experiencing life and interacting with people. Both of those lifestyle choices are going to include pain, but only one of them will also include depression."

I like to share this story because I think that all too often we ascribe purely physical explanations to our suffering. Sure, there may be an organic origin to something, but the behavioral choices we make can and do have an impact. We choose how we cope, and our choices in turn influence our condition. With chronic pain, we fall into a cycle of pain that leads to a depression that amplifies the pain and leads to more depression that further amplifies the pain and leads to more . . . you get the idea. Laughter may not be a cure for anything, but a little positive affect and momentary pain relief sure can help interrupt that cycle.

Norman Cousins was onto something.

LAUGHTER REDUCES STRESS

I MENTIONED EARLIER THAT MY FAVORITE BENEFIT OF LAUGHTER IS that it makes us happy. This is true. I love laughing primarily because it makes me happy. I love making other people laugh because it makes them happy. It makes me happy to make them happy. I am such a giver. I should get a medal.

Happiness may be my favorite outcome of laughter, but it is far from being the most important. That honor belongs to the fact that laughter helps reduce stress. It does, it so does. We can live without being happy, but we will have a really tough time living under prolonged stress.

The negative impact of prolonged stress is wide reaching and can contribute to a variety of mental and physical disorders. Seriously, you can pick up the *Diagnostic and Statistical Manual* (DSM) of the American Psychiatric Association (if you are unfamiliar, this is the book that defines all of our psychological diagnoses) and flip to any random disorder, and the chances are really good that the disorder you've selected is either partly caused or significantly influenced by stress. Stress is a component in nearly everything that commonly goes wrong with a brain. It is a factor in the development of substance use disorder or addiction. Stress is a big part of generalized anxiety disorder, panic disorder, and the development of phobias (irrational fears). It increases our risk of developing obsessive-compulsive disorder (OCD), and many of the personality disorders, including borderline personality disorder (BPD) and antisocial personality disorder (the thing that psychopaths have). It is also the basis of post-traumatic stress disorder (PTSD), and it is associated with the onset of schizophrenia—and early exposure to stress may even be one of the causes. And of course stress is a major predictor of depression.

Stress, however, does not just lead to mental disorders. Even in the absence of an extreme condition, it can cause mood swings, difficulty concentrating and headaches, and loss of sleep and lethargy. Also, it can kind of make you act like an asshole.

Physically, stress takes its toll on the body with increased blood pressure and heart rate, a reduction of the immune system, lack of sex drive, and various aches and pains. Chronic stress can contribute to heart disease, diabetes, obesity, asthma, and even cancer.

A little stress is okay, and sometimes even beneficial to us. I previously described how whenever our brain encounters a stimulus, one of the first things it has to do is determine if that stimulus is going to kill us. Well, if the stimulus does represent a threat, then it is in our best interest to do something about it, so our brain initiates a set of physiological changes, such as increased heart rate and tightening of our stomach muscles, which get our body ready for action. Those changes encourage our "fight or flight" response, and are extremely helpful if we have something to fight or run from.

For example, if we are out in the woods and we see a giant grizzly bear in the distance, we may want to ready our body for some form of action. Suppose that bear notices us and the fact that we happen to be carrying a large open container of smoked salmon (maybe a poor choice given the context) and the recent issue of *Bear Playboy* (you know, for the articles—by the way, unless you really want to confuse your brain never ever Google "Bear Playboy"), we might need our heart to pump a little faster and start fueling the muscles. Suppose he sees the "Fuck

Bears" sticker George Clooney put on our gear that seemed so funny at the time. Suppose all this encourages the bear to run toward us. Being attacked by a hungry, horny, and pissed off grizzly bear would probably be a really good situation for some old-fashioned fight or flight (although I understand you should not try either with a real grizzly if survival is at all important to you), but very few of us are going to be attacked by bears. In fact, we live such awesome and comfortable lives that a lot of us will probably never face a truly life-threatening situation. We get stressed over an exam, a job, a Facebook post, a book deadline (I swear I am typing as fast as I can!), etc. . . . Our stressors typically aren't life threatening, but despite the fact that we have all sorts of various types of stressors, we only have one stress response. Most situations we face do not need to be dealt with like a bear attack. And that is part of the problem.

If there is one redeeming aspect of being attacked by a bear, it is that the experience probably is not going to last very long. Our brain recognizes that our body is in trouble and initiates some physiological changes to increase our likelihood of our short-term survival at the cost of our long-term objectives. One way or another that danger is going to be over fairly quickly, and so will our stress response. However, some of the more common stressors that we experience in our lives, like job stress, have the potential to really linger.

A lot of our stress is not even rooted in the outside world but instead comes from inside our own skull. Our thoughts are our biggest source of stress. Negative or anxious thoughts, influenced by beliefs, can be interpreted by our brain as a threat and activate our stress response. We might call these thoughts guilt, regret, spite, blame, or maybe even worry. That's right; I am talking about worry (among other things). Sometimes thoughts don't just go away and we ruminate on them for a period of time. It's enough to drive someone crazy, literally, as it is a major factor in the development of emotional disorders. You might recall that in my introduction I implied that we need to start understanding that negative thoughts are part of the cause, not the symptom, of a disorder. As I stated earlier, it isn't the case that you worry so much because of your anxiety disorder, but you have that disorder because you worry so much (to put it very simply). We all do this. Some just do it more often than others.

I have already shared that I am a very happy person, so it should not be surprising when I say that I do not ruminate on negative thoughts very often, but I do, of course, sometimes do this. I am human after all and am prone to entertain the occasional nasty thought as it pops into my head. For example, one day recently my girlfriend, Sarah, sent me a text message and I replied with a question, expecting her to reply. She did not, which is unusual for her. Normally she replies within a few minutes, but this time an hour had passed since I had last heard from her. The realization that she had not replied became a stimulus that triggered thoughts as I began to imagine the many possible reasons why she may not have answered me. Was she angry with me? Did something happen to her? Was she okay? Just asking myself those questions led me down a very dark path and I imagined all sorts of horrible scenarios that could have been preventing her from texting me. From simply deciding that she no longer wanted to be in my life to the possibility of her being involved in a car accident, my head was full of negativity that I could not shake. So yeah, not every worrier has a disorder. It happens to all of us. For whatever reason, my normally optimistic brain had gone off the deep end that day and I ruminated on that negativity. My heart rate increased, I had a hard time focusing on anything else, and became unusually anxious until she finally texted me back. Thankfully, she'd just fallen asleep.

Ruminating negative thoughts and long-term stressors activate our stress response just like the charging bear, but unlike with the bear, the stimulus stays with us. Therefore the associated physiological changes we experience are prolonged for an unhealthy duration. The mental and physical problems we attribute to stress come from prolonged exposure to those changes. Acute stress can be helpful, but chronic stress is a very big deal.

Prolonged stress can be so extremely detrimental to our health that it is for this reason I believe that the primary purpose of laughter is to reduce stress and calm us down. Sure, laughter makes us happy, provides us with pleasure while reducing pain and doing other stuff I haven't mentioned yet, but the main benefit we get from laughter is stress reduction. It is nature's built-in stress management tool. We need stress, we need to feel anxious or aggressive when we are threatened, but we also need a means to calm ourselves down when that threat has been reassessed. You will recall that I previously mentioned that according to relief theory, humor is often used to relieve tension. It is also often used as a defense mechanism against painful thoughts and emotions. Both of those are more simply stated as ways to reduce stress. We can use laughter to reduce the stress associated with negative energy (relief theory) or use it to cope with the stress associated with painful memories or thoughts as a defense. Both ways, and more, are stress management.

This is why benign violation theory works so well to explain what we find funny. With BVT, we experience a normally threatening stimulus but then reappraise it as benign, resulting in a perception of humor.

When Sarah finally did text me back, I called her and told her how I had been worrying so much. We both laughed at the situation, immediately helping to reduce my stress level.

Laughter is nature's stress management tool, so how does it accomplish this? Well, one of our body's reactions to a perceived threat (stress) is to stimulate the release of a hormone called cortisol from our adrenal glands. Cortisol, often referred to as the stress hormone,[66] is then involved in producing many of the physiological responses we associate with stress. For example, it acts on the inner lining of our blood vessels, the endothelium, causing our blood vessels to contract. This vasoconstriction therefore elevates our blood pressure.[67] Conversely, one of the physiological effects of laughter is that it lowers our blood pressure. When we laugh, we engage in a lot of bodily activity and the muscles responsible for it all need oxygen. Our blood vessels relax and expand to allow for better circulation and vasodilation.[68] Our blood pressure drops as a result.

The stress hormone cortisol has a number of effects on our physiology, many of which are related to our stress response, keeping us at a heightened state of arousal. In contrast, when we laugh there is a lot of activity in our body and our brain. One of the physiological effects of laughter is that it significantly reduces cortisol production, bringing our body back to a calm state.[69] This effect is observed as we are laughing and then continues for some time afterward. Self-reported feelings of stress also decrease the more an individual laughs.[70] Other stress-related hormones, such as epinephrine, are also reduced through laughter.[71] Epinephrine in particular is implicated in hypertension and heart disease, implying then that a good hearty laugh is good for the heart. Okay, I'll admit that isn't a "laugh out loud" kind of joke.

In my introduction I mentioned that I began performing stand-up comedy in San Francisco, originally by taking a class at the San Francisco Comedy College. I had been wanting to get into comedy for a long time before that, but what really pushed me into the doors of that building was stress. San Francisco is a very expensive city to live in, especially if you made the mistake of not being from there or not buying a home there thirty years ago. To make ends meet, I took a position that the human resources person would later describe as a very bad match for my training, skills, and personality type. The job was in a very stressful industry, and I worked extremely long hours in an office with very few coworkers. I wasn't right for that job,

I hated that job, and, without my realizing, that job made me miserable.

One night I was out with my friend Joanne, also a psychologist, in downtown San Francisco when we stumbled onto Comedy College together. The sign read "classes starting soon" or something like that, and she reminded me how I had always said I wanted to try stand-up comedy. My first reaction was that I would never be able to fit it into my work schedule, but it planted the seed of interest. A few stressful weeks of discontentment later, I realized that I needed something, anything, to help me relax after work. After a particularly bad day, I called the college and signed up for their next available opening.

I remember that I almost missed that class due to having a hard time getting out of the office that day, but as soon as it started I began to relax. I felt myself start to loosen up, places where I didn't even realize I had been tense, and clear my mind. Comedy, and laughter, became my outlet. I left class excited about what I had begun and slept better that night than I had for a long time. In the morning, I woke up still feeling positive and went into my miserable office. It was still terrible, but for some reason seemed a little more tolerable.

The more I practiced stand-up and the more comedy shows I attended, the more I felt myself come out of the stress-induced misery that I had been experiencing. Comedy had awoken me from a prolonged stressful state, and probably a mild case of depression. Eventually, my interest in comedy overcame my desire to continue punching in at a job I despised and I happily parted ways with that company. In fact, I decided to leave the entire industry completely and try my luck with a major career change. Life has had its ups and downs, but it has been an awesome ride ever since.

Looking back on that experience, I give comedy a lot of credit for helping me de-stress and cope with a difficult period of my life. However, it did not have to be comedy. When I was younger, I was an artist, a poet, an occasional singer, even a radio DJ. I worked as a doorman at a swing jazz club in New Orleans on weekends when in graduate school and I've been a prolific photographer. Throughout my life I had always had some creative outlet to balance out whatever full-time drudgery occupied my days, but for that year in San Francisco I had none. Thankfully, Joanne and I saw that sign one night.

Laughter is not the only activity that can help reduce cortisol levels. I suspect that some of the activities I've listed as my former creative outlets might help lower stress hormones as well. However massage,[72] listening to music,[73] and regular tango dancing[74] have all been shown in empirical research to help lower cortisol and in turn help manage stress. It makes sense that there would be experimental research demonstrating the effects of massage; however, I find it interesting that there is also a substantial amount of research on the stress reducing benefits of tango. Coincidentally, my girlfriend, Sarah, is an accomplished tango dancer and, lucky for me, also a gifted masseuse. Lucky for her, I am good at making the humor. Together, we manage stress remarkably well.

Stress management is extremely important to our health, and as I've stated I see laughter as nature's stress management tool. It has plenty of other benefits, but this view seems consistent with our experiences and even observations that have been made about the ways in which other species use laughter (or related behaviors).

There may be one more benefit of humor and laughter related to stress. I believe that laughter helps us develop a higher tolerance to stress. We all encounter stressful stimuli and we all experience stress. However, we differ on the level of stimulus intensity we react to. Our stress tolerance refers to how much we can handle before our stress response kicks in. We all vary on this. Some of us get really worked up after being exposed to a low intensity stressor. I think of traffic as a pretty low intensity stressor; in fact I hardly ever get stressed out in traffic. To me, traffic is just cars moving slowly. Cars are safer when they are moving slowly! What could be so stressful about that? Sure, traffic sucks some times, but I rarely find it stressful. And yet for some people traffic is a major source of stress. They become anxious, agitated, and even enraged. I have known many people who are prone to bouts of road rage. Someone moves their car close to theirs or "cuts them off" and they get so angry they'll honk, tailgate, and even follow the transgressor for miles. They start taking some major risks. The road rager will literally endanger their own life because someone forgot to use a turn signal. I could never understand that behavior from a rational standpoint, but that's the key. It isn't rational. Their brains have interpreted that other driver's behavior as a potential threat, and their fight or flight response is now at the wheel.

On the other hand, there are people who display an extremely high tolerance for stress. These are people who can remain calm under some extraordinarily intense situations. I always think of people with extreme military training as people with a very high tolerance for stress. Members of elite Special Forces trained to jump out of perfectly good airplanes into combat zones, for example. I also suspect that professional hit men, not that I have ever known any, are highly stress tolerant. Their "job" would no doubt require them to remain calm under enormous pressure. Psychopaths definitely have a very high stress tolerance.[75] That is basically the defining trait of a psychopath: not having a normal physiological reaction to thoughts and behaviors that produce anxiety in the rest of us. Psychopaths, for whatever reason, are naturally stress tolerant. Whether it is dispositional or the result of having received extensive training, or some combination of both, there are some people who have the ability to suppress their stress response when most of us would be losing our damn minds.

Our level of stress tolerance has a lot to do with our sense of control. There has been quite a bit of research that shows that the more we feel in control, the less reactive we are when threatened.[76] By feeling in control, I mean that we feel as if we have the ability to influence the situation either by solving it or escaping from it. It is a cognitive reaction to encountering a stressor, as opposed to an unconscious "gut" reaction. When we feel as if we are in control of a situation, our prefrontal cortex sends a signal downward to shut off our stress response.[77] We will probably engage in some form of action, but we get to remain calm while doing so. Stressful situations, from our brain's perspective, are just problems that need to be solved. Therefore it should be no surprise that research shows that we can develop our sense of control by solving problems.[78] Every situation, no matter how minor, that we successfully deal with helps contribute to developing our sense of control. In other words, the more we overcome, the more likely we will remain calm in the face of the next potential crisis. This is why training can increase our stress tolerance.

I believe humor also helps us develop our sense of control, and therefore helps increase our tolerance for stress. Again, if we think about our recognition of humor in terms of reappraising a potential threat as something benign, then every time we find humor in life we are engaging in a process very similar to solving a problem. Humor is the solution, laughing is the outcome. Obviously there are real problems that we must face, but the more we are able to recognize them as nonthreatening, the more likely we will do so when we encounter the next problem. With a higher tolerance for stress, we will find ourselves getting stressed a lot less often. Humor therefore does not just help reduce the direct effects of stress in our lives, but also how frequently it is an issue.

I am unsure if anyone has been able to quantify this, and given the variety of stressors in the world and our own individual experiences, I have a hard time imagining how someone would. However, I know that I seem to have a higher tolerance for stress than most people. Now, I am definitely no psychopath, so please don't get scared, but I do have a high tolerance. I would not be able to live the life I lead, touring all over the country and dealing with different challenges every day, if I were more easily stressed. Throughout my life, I can remember situations in which I remained calm while others around me were stressing, but I never knew that I had a higher tolerance than other people until I was in graduate school. Graduate school is stressful and if you have been, you know exactly the kind of sheer, sleep-depriving, anxiety-provoking hell that it can be. It is torture *that we pay for* and is just awful. I don't recommend graduate school to anyone, although clearly there are some rewards or I would not be writing this statement. Anyway, my story takes place in graduate school but it is not about it.

If you were paying attention earlier, you may have picked up that I spent part of my education in the city of New Orleans, Louisiana. I love New Orleans and always have since my first visit. In my opinion it is one of the greatest cities in the world, and I go there as often as I can. New Orleans is a city with a unique geography, an incredible history, and a charming culture. I lived in the French Quarter district. If you have never been to New Orleans, the Quarter is the area that most tourists visit. It is full of old architecture, cobblestone streets, and plenty of restaurants and bars. If you have been to New Orleans, you are probably wondering how I managed to finish graduate school at all. Sometimes I wonder that myself. Bourbon Street runs through the Quarter and is one of the most bar-dense areas I have ever seen. They have alcohol sales twenty-four hours a day in New Orleans, and you can legally drink on the streets. I lived about two blocks off of Bourbon Street.

New Orleans is also a high-crime city, and it has been for a long time. At the time I lived there, the French Quarter was particularly crime-ridden although I have heard things may have cleaned up a bit in the post-Katrina era. I lived in a high-crime neighborhood, in one of the high-crimeiest cities in the country. That never bothered me; it was just part of the price of living in such a fantastic slice of the world.

One day I was walking home from the grocery store with my roommate. The store was only a few blocks away from where we lived, and we both were carrying bags of groceries in each of our hands. It was the middle of the day, nobody was on our street, and just as we were right outside of our apartment a guy holding a gun popped out from behind a parked car and said, "Give me your money"—or something to that extent. We were being robbed, in broad daylight, right in front of our apartment. By the way, I feel like if you make it home that's "base." Like, he should have robbed me down the block. That just seems fair. I didn't tell him that we were in front of my apartment, or things could have gotten way worse.

So I was being held up at gunpoint right there on the street. It was the first time that something like this had ever happened to me and it is important to appreciate that because I had no prior experience with a situation like this. I had no idea how I would react in such an extremely stressful situation. My roommate understably reacted like I think a lot of people would. He became overwhelmed with stress and anxiety, completely flushed, and started to perspire. You could see the anxiety and fear in his face. He began to stutter; his speech and his hands were shaking. In fact, he dropped his bags of groceries and with his shaking hands he nervously reached into his pocket to pull out his wallet and hand it over to the guy. He was really just overwhelmed with emotion and I remember looking at my roommate and thinking, *Wow, he is really freaking out.*

Surprisingly, I remained completely calm, very calm. In fact, I remained so calm that I started to negotiate with my mugger. I reasoned with him and said, "Sure, you can have my money. That makes sense, you've got a gun and you are holding it out in the middle of a city street. You've taken a substantial risk here, and I feel like that risk should pay off for you." (I had recently attended a seminar on investment.)

I went on to explain to him that he could have my money, with no objections or argument, if he agreed to let me keep my wallet. My wallet had my driver's license in it, my car registration, student ID card, my credit cards, which would be of no use to him five minutes after he left, and other important documents that would be a serious pain in the ass if I had to replace them. I said, "If you agree to let me keep my stuff, you can have my money. Those are my terms."

Now here is the thing about armed robberies: it is a very stressful experience to be held up at gunpoint. Seriously, take my word for it and don't try this at home. I would imagine it is also stressful to hold somebody up at gunpoint. My mugger seemed as if he was also pretty stressed out. His hand was shaking as he continued to point the gun at me, repeatedly saying, "Just give me the money; just give me your wallet."

"Okay, calm down. Just relax," I went on, "You can have my money, there is no argument there. I just want you to agree to let me keep my stuff." We went back and forth like this about three or four times.

Eventually he calmed down. He said, "That sounds reasonable." So I pulled my wallet out, and right in front of him I took out the little bit of money I had (only about $12, we are talking broke graduate student here) and handed him the cash. Then he said to me, "Can I look inside your wallet to make sure there is no more money in there?"

I thought about it and figured that made sense. I could have hid some larger bills behind my driver's license or tucked some away in a fold. Before I gave it to him, I said, "Look, I'll let you go through my wallet but you have to promise that you will give it back. You have to promise you are not going to run away with it."

"I promise."

So I hand him my wallet and he starts flipping through it; checking every nook and cranny. When he was satisfied there was no more money in it, he gave it back to me. I said, "Thank you." He said, "You're welcome."

And then I asked if my roommate could get his wallet back too. He agreed and took the money out of my roommate's wallet before handing it back to him. My roommate said, "Thank you" and the mugger said, "You're welcome."

And then our mugger said, "Well, I guess I'll be going now."

"Just leave man, it's not like we are friends now, you are ROBBING us. Just get out of here."

And that is the story of how I realized I have a much higher tolerance for stress than most people.

Just because I know that my mother is reading this book, I want to make sure I am very clear about something. Nothing in my wallet was worth risking my life over. Well, except that picture of Mom.

I also want to make something else perfectly clear. I was not in control of that situation. I was not trained in self-defense or planning on dodging a bullet or anything ridiculous like that. Assuming the gun was loaded, if that man wanted to shoot me at any time, I was going to get shot. Objectively, I had absolutely no control over the outcome of that situation, but my brain reacted with a sense of control. My brain reacted as if it knew that there was something I could do to influence the outcome, and in turn I remained calm so that I could do just that. By not engaging my stress response, my rational thinking remained intact and I was able to give the situation some thought and come up with a mutually beneficial option. In contrast, my fully stressed roommate had limited ability to consider his options and was barely able to hand over his wallet in compliance. The reason I just shared that story with you is this: we don't actually have to be in control of a situation, we just have to feel as if we are.

I have no idea how I developed such a high tolerance for stress. I was not a psychologist as I was growing up. I was just, you know, growing up. I wish I knew, because as I shared earlier, stress is a huge factor in so many of the mental and physical problems that we face. If I had some insight on how to develop a high sense of control or raise children who will grow into adults who manage stress well, I would be writing a very different book. I'd say, "Hey, everyone, raise your kids just like this," and sit back and watch my new bestseller take off while partying with Clooney and the guys from *Jackass*. I am sorry that I do not have that insight to share, but I have a few ideas.

When I was growing up my father was in the military, which meant that my family moved around a lot. About every four years I was in a new school in a new state and surrounded by new people. This presented a great big set of potentially solvable problems for my developing brain, and I tackled each of them every time. Most of the time I was successful, a few times I was not, but overall that problem solving gave my brain plenty of practice. As I suggested, developing a sense of humor probably helped as well by training my brain to reassess potential threats. Sometimes my sense of humor was part of the solution; it certainly helped me to make new friends and adjust to my new surroundings with every move. Somewhere along the way, I also learned that I could use humor to help defuse tense situations and I talked my way out of a whole bunch of ass kickings. This is just speculation here, but maybe all that practice in problem solving and reducing tension somehow convinced my brain that it could do the same thing with a gun pointed at my face. And maybe that is exactly what I did.

OTHER THAN THE OBVIOUS, WHY IS STRESS MANAGEMENT SO IMPORTANT?

S TRESS IS OUR RESPONSE TO SOME PERCEIVED THREAT AND KICKS in a whole host of physiological changes that ready our body for immediate action. Short-term stress has a purpose in motivating us to act and helping to keep us safe. However, prolonged exposure to stress is a major contributor to depression, anxiety, hypertension, obesity, and a plethora of other mental and physical problems. Laughter helps reduce stress and calm us. At the same time, laughter makes us happy. This is not a coincidence; stress and happiness are directly related. Stress is the opposite of happiness. We may not always define them like that, but physiologically speaking we cannot experience happiness when we are under stress. It would be like trying to sleep while we are awake. Or like trying to eat when you are full . . . never mind, we seem to be capable of that one.

I previously discussed how happier people live longer than less happy people. Did you even wonder why this is? So far, all I have shared with you is some research about smiling dead baseball players. I gave you no explanation for why the smiling players outlived the others by seven years. We just kind of passed over that point, didn't we?

There is no simple answer of course, but the happiness/ stress dichotomy definitely plays a role. Many of the stress-related illnesses that we suffer are going to take their toll by shaving years off of our lives and most definitely making us less happy. If we are generally unhappy, we are inadvertently setting ourselves up for a higher risk of illness. Happiness is healthy — even the US surgeon general is discussing the importance of happiness—and this makes a lot of sense when you consider that happiness occurs when we are not under stress. I believe that it is not so much the case that happiness is healthy, but that happiness is indicative of the absence of stress. Therefore I think it is an easier and more practical goal to focus on being happy than to focus on eliminating stress.

Comedian George Burns said, "If you ask me what is the single most important key to longevity, I would have to say it is avoiding worry, stress, and tension. And, if you didn't ask me, I'd still have to say it." George lived to be a hundred years old.

Laughter helps reduce the impact of stress on our lives. Developing a good sense of humor can help increase our tolerance to stress and allow us to experience stress less frequently. Given all we know about the value of happiness and its effect on our health and longevity, obviously stress reduction should be an important concern. So let's all learn to laugh it off and quit taking life so seriously. We need to learn to laugh more often.

LAUGHTER IS HEALTHY

I CONSIDER MYSELF VERY LUCKY THAT UP UNTIL ABOUT THREE years ago I had three surviving grandparents. Longevity apparently runs in my family. If I play my cards right I might be working comedy clubs and giving talks for a long, long time. That means another few decades of living out of a suitcase, bouncing from hotel to hotel, and driving for hours a day just to entertain a casino full of hard-to-please drunken truckers in Winnemucca, Nevada. Wow, longevity might be overrated. I'm kidding; I love life on the road. No, really, I do.

Anyway, in 2012 both of my grandmothers were alive and doing well. My only surviving grandfather was sick, really sick. In fact, he was sick for a very long time. As far as I could remember, he dealt with health issues. When he did work, it was in construction; he had been a professional welder for I don't know how long. During that time, he swallowed a lot of asbestos, because apparently that's what they did on their breaks. With multiple cancers and other issues, his health was poor. He was in pain for pretty much all of my life, which was almost half of his. For the last few decades of his life, he rarely left the house. Then for his last few years, he grew very frail and was entirely dependent on my grandmother, who became his caretaker. He finally lost his battle in 2012 at the age of eighty-nine.

One thing about my grandfather, regardless of his physical condition, is that he was always in really good spirits. He had a fantastic sense of humor and laughed a lot. His laugh, even when it was a struggle for him to produce it, was infectious and he would often smile through his set of false teeth. He was also a huge fan of cartoons, especially the classic slapstick comedy shows like *Looney Tunes* and *Tom and Jerry*. Every day, as part of his routine, he would wake up and make his way into the living room, sit in his favorite chair, and flip on his cartoons. There he would sit, laughing his ass off for the next hour or so. He loved those cartoons!

My grandmother, on the other hand, did not care for them too much. When I would visit, she would often complain to me that "he is in there, watching those stupid programs." She didn't understand why he would waste his time like that. I didn't understand it at the time either, none of us did, but I bet those cartoons were helping him manage his illness. In a way, it seemed they were animated life support. Those cartoons were keeping him alive.

Laughter can help prevent and slow the progress of cancer. Shit, it can totally do that.

Forgive my harsh language, but I feel like that is a pretty bold claim and has to be followed with an expletive.

Cancer is a horrible disease that we still do not completely understand. For a variety of possible reasons, cells in our body begin to reproduce abnormally in the form of tumors. These tumor cells can then spread to other parts of the body. There are many known factors that contribute to its development. This list of course includes the voluntary use of substances like tobacco and exposure to environmental carcinogens like the asbestos that affected my grandfather. Obesity and the lifestyle choices that contribute to weight gain, poor diet and low levels of physical activity, also increase cancer risk. It can be caused by some infections, such as human papillomavirus (HPV), and some genetic factors. However, in addition cancer is also influenced by internal factors including our affective state or mood, stress exposure, and cognitive activities such as negative thinking.

The body has its own natural defenses against cancer. These include special cells in our immune system, which go by names like gamma-interferon cells, T cells, and natural killer (NK) cells, and have been shown to be increased by laughter. A demonstration by Bennett and associates (2003)[79] showed that the more time a patient laughed in a specific time period, the greater the increase in the activity of their NK cells. Laughter, and perhaps the positive mood which it inspires, helps our body fight off cancer.

Now, you might be thinking to yourself, *That's great, Brian, but how is someone supposed to laugh when they have a disease like cancer?* I can't answer that, mainly because I am not in the room with you, but mostly because I have never had to experience the horrors of a disease like cancer. However, I do know people who have and somehow they manage to find humor in dealing with this difficult illness.

My friend and fellow comedian Rudy Martinez and his wife, Samantha, have been dealing with cervical cancer ever since she was first diagnosed three years ago. I asked Rudy how he and his wife find humor in cancer. He said, "At the most inappropriate times . . . like when my wife's legs are in stirrups and the doc is elbow deep in her." Rudy and his wife treat cancer like any other subject, and freely joke about it. As he put it, humor defanged the beast. At one of my seminars in Florida, I invited Rudy to share his experiences with the audience and he told the following story:

> At first it freaked me out, and we go in for her first doctor's appointment. They have her in the stirrups and my wife is in pain. She's squeezing my hand and she's crying and all I can think is, *I need to be anyone other than me right now. I need to make sure I say the right thing.* But

I am not anyone other than me because I am who I am. She's squeezing my hand and she says to me, "I'm so sorry." And I start thinking to myself, *Say the right thing, say the right thing, say the right thing . . .* ". . . Honey you have nothing to be sorry about!" and she says "no, I do! We're not going to get to have sex for a long time!" *Say the right thing, say the right thing, say the right thing . . .* I look at her and I say, "No honey, YOU'RE not going to get to have sex for a long time."

Did he say the right thing? I don't know. However, I do know that his comment made his wife laugh in spite of the uncomfortable situation she was in, and they have continued to laugh together throughout her treatment. According to Rudy, for him and his wife, "Laughter is a big fat middle finger at cancer." And with every laugh, those natural killer cells are becoming more active.

The cell types I mentioned earlier, NK cells and such, do not just help protect the body against cancer but are also involved in protection against other illnesses as well. Laughter gives a good general boost to our immune system and even reduces the frequency of colds. Seriously, we get sick a lot less the more we laugh.

I think of myself as a pretty good example of this. Without knowing what my cell counts are, I can say that I seem to have a relatively strong immune system. Every day when I am on tour, I spend my time in a room full of several hundred germ-infested disease hosts, many of whom cough and wheeze throughout the day, and yet I rarely get sick. I occasionally lose my voice because of the strains of speaking for a living, but I rarely catch a cold or flu. This could be luck, and that probably has a lot to do with it, but I am positive that my frequency of laughter and stress management help keep me healthy. I can tell you with certainty that it is definitely not related to any medication, nutritional supplements, or any over-the-counter placebos. You may have noticed, because I've said it, that I attribute at least some of my health to stress management. Stress diminishes our body's ability to protect itself from infection. Specifically, the cortisol our body releases when stressed inhibits the production of gamma interferon cells and T cells. In the previous chapter, I discussed the stress response as a set of physiological reactions intended to favor our immediate, short-term survival at the expense of more distal, long-term goals. Protecting the body against illness is a long-term concern. What does health matter to your survival if you are just going to be mauled to death by an angry grizzly bear? Therefore when we are under stress, our bodies suppress our immune system. This, again, is another reason we need to manage our stress.

Centuries ago, some anonymous genius said that "laughter is the best medicine" and that phrase sticks with us because it rings true. By boosting our immune system in multiple ways, laughter is great preventative medicine. It also helps us heal and recover from illness.

An old Irish proverb goes, "A good laugh and a long sleep are the two best cures for anything."

But the impact of laughter on improving our health is not only a function of increased activity in our immune system. Because laughter helps reduce stress, it helps reduce the impact of stress-related illness. Laughter can help reduce high blood pressure and our chances of suffering a heart attack, and can help prevent the development of type 2 diabetes.

If we do develop type 2 diabetes, laughter can help us manage the illness. It has been shown that a Norman Cousins–style daily dose of comedy can reduce blood sugar and increase our tolerance for glucose. It should be noted that while this benefit may be of value to those managing diabetes, this effect is observed in non-diabetics as well. In other words, laughter can help anyone lower their blood sugar.

For these reasons, I like to state that laughter can help our body compensate for poor lifestyle choices. Smoking, overindulging in food, and not exercising enough are not healthy ways to live our lives and eventually may ultimately cause us a great deal of harm. However, laughter and positive affect can help offset those choices. It won't make us immune to the negative consequences of unhealthy choices, but it can help delay them. As I mentioned in my introduction, I have been overweight my entire life, yet my blood sugar remains low (I know this because my diabetic mother insists on pricking my fingers every time I visit her) and my blood pressure is in a relatively healthy range (although at the high end). Poor lifestyle choices will eventually catch up to me, I am sure, but the fact that they have not yet could be partly attributed to the health benefits of laughter.

Discussing how laughter may help compensate for unhealthy lifestyle choices makes me think of Rolling Stones guitarist Keith Richards, a man that Robin Williams once said "is the only man who can make the Osbournes look fucking Amish." Keith's smoking, alcohol consumption, and drug use are the stuff of legend and the fact that he, now in his early seventies, is still among the living is an easy punch line for any hack comedian. I'm no judge of how long someone should be expected to live given a chosen lifestyle, but I appreciate the fact that besides being an amazing guitar player, Keith has been described as having a fantastic sense of humor. In the book *Keith Richards on Keith Richards* (2013), he says, ". . . one of the most precious things in this world is sense of humor." He is so well known for his unique laugh that there are thousands of YouTube videos devoted to showcasing what *GQ* magazine once listed as the "Most Ear Catching Laugh" (2011),[80] of all time even! Now, I have never met him, and all I know about him comes from public sources, but I do find it interesting that a man known for hard living and living long also has a great sense of humor and laughs a lot. I'm sure being a stinking rich rock star hasn't hurt his survival chances either.

So yes, I think there is something about laughter and positive affect that help to offset unhealthy lifestyle choices. However, a lifetime of laughter does not guarantee immunity from disease. Comedians as a whole do not have a reputation of taking good care of themselves; many of them famously abuse alcohol and other drugs, and a lot of them have some giant appetites. This is, of course, a sweeping generalization, and there are definitely healthy comedians. George Burns may have smoked cigars but he also exercised every day and lived to be one hundred. Still, it has always seemed to me that some issues disproportionately affect comedians more than the general population. Some of my favorite comedians are tragically unhealthy. For example, Gabriel Iglesias, otherwise known as Fluffy, whom I discussed earlier in this book was diagnosed with type 2 diabetes in 2012 and given two years to live. This prompted him to lose more than one hundred pounds, so hopefully he will be around much longer than that.

Finally, the act of laughing even helps defend against respiratory infections by circulating the air in our lungs faster. So even with the boost to the immune system, stress reduction, and blood sugar reduction, there is that. That is a tremendous list of benefits, and those are just the physical ones.

Unfortunately, we can still get sick, and depending on the nature of our illness, sometimes it is hard to remain optimistic. Sometimes, we have to come to terms with our condition. In research, positive psychologists have demonstrated the importance of not only optimism, but also acceptance, to our happiness. Sometimes, the best thing we can do for ourselves is accept that it is what it is. Having a sense of humor can facilitate this.

There is plenty of research that shows that attitude and optimism can go a long way in helping us heal. It isn't as simple as "mind over matter" as we cannot will ourselves to recovery, but our attitude does impact our immune system. On my speaking tours I interact with a lot of nurses who work in various hospital settings. Several have told me that they can often predict a patient's outcome just by their disposition when admitted. Those who seem happier and more optimistic tend to fare much better than those who seem to have given up.

Even before knowing anything about the physiology behind it, this made perfect sense to me. The brain and all the other organs of the body comprise the human being. It always seemed that a being who wants to live or has something to live for will try hard to do just that.

I am acquainted with a comedian from Oklahoma named Stan Silliman. He performs all over the middle of the country but doesn't stray too far from home because he is being treated for cancer. In the sixteen years since his diagnosis, he has undergone six operations, had twenty-four MRI scans, and spent eight-and-a-half hours in radiation. Throughout his ordeal, humor has helped him remain resilient. Laughing and making others laugh (particularly his wife) have been instrumental to helping him cope. He told me:

> Laughter is a good medicine. It lifts my spirits. But even more lifting, to me, is the ability to make someone else laugh. And I have an ideal audience, my wife of forty-seven years. She has the most fabulous dazzling smile, an extremely [hearty] laugh and it's a total pleasure if I can get her to laugh. But she is a tough crowd. Her standards are incredibly high. I figure if I get her to laugh at least three or four times a day, I've done a good job, maybe yeoman's work. Not only does it give me pleasure to see her beautiful teeth and laughing hard but it has sharpened my humor coming up with ways to get her into laughing fits.

Stan has been very prolific. In the time since his diagnosis, he has managed to author eight humor books and maintains a humor website called Silliman On Sports (www.sillimanonsports.com). He performs in clubs, and regularly entertains hospital staff and other patients on his visits. To those dealing with a similar diagnosis, he says to remain optimistic.

Cancer is not nearly the death sentence it once was. It can be beat or if not completely, weakened and detained. Work on a plan with your doctors and don't be afraid to get second or third opinions. Plus if you run into an oncologist who is a sourpuss or has bad bedside manner, keep running. Get yourself another one. Research about your cancer as much as you can. Know your options and *try to have fun when the opportunity occurs*. Keep your spirit up and if you can't find enough occasions to laugh, schedule some.

Laughter boosts our immune system and helps prevent certain types of disease, but remember that it is not a cure for anything. It helps. It makes life better, but all it really "cures" is silence. As my friend and fellow comedian Dave DeLuca says, "To those people who say laugher is the best medicine, I'd like to tell you about penicillin."

COMEDIANS WITH DIABETES ON COPING WITH DIABETES

I N PREPARATION FOR THIS BOOK, I REACHED OUT TO MY NETWORK of comedians and asked if anyone would like to share their perspectives on certain health related topics. The results were overwhelming, but this was not surprising. Comedians in general are extremely nice people and they absolutely love talking about themselves (there is a reason why so many jokes start with the phrase "a little about me . . ."). Anyway, I am very appreciative of all of the responses I received. For anyone dealing with a new diabetes diagnosis or struggling with the hassles of managing the illness, perhaps some of these responses will be of assistance.

How did you cope with being diagnosed with diabetes?
"That's tougher to answer than I thought. I was twenty-five when I was diagnosed so my life was pretty all over the place as opposed to being an adolescent, like most type 1s. It was hard enough . . . managing life [as] a twenty-something . . . [before] adding diabetes management into the mix. The hardest part that I realized later on is not having a network of people to talk to. I was pretty isolated and that caused me to make a lot of mistakes; mistakes that I didn't correct until years later."
—Chelcie Rice (Atlanta, Georgia)

"Not well. I had a very emotional reaction to the diagnosis. When I developed diabetes, I had been working out for about two months, hitting the gym every day. So I didn't notice that how much water I was drinking was unusual. I was filling up a gallon jug once or twice a day at least, and urinating much more frequently. *That makes sense*, I thought, *I'm drinking all this water because I'm working out and losing weight!* When I got the diagnosis after a physical, I looked up the symptoms: unexplainable weight loss, unquenchable thirst, and frequent urination. Getting the news was devastating in the first place, but to find out that it wasn't even my efforts at the gym that were causing the weight loss. I became depressed, and the girlfriend I had at the time was rude about it. 'What did you expect? You're fat.'"

—Jason Ward (Boise, Idaho)

"I was truly shocked! I was in a particularly bad place with my life and had been living with my former partner's parents in Riverside County. I was definitely also dealing with some situation-based depression. I had been feeling pretty sick for about two to three months. As it got closer to when I finally went into a hospital, my vision was blurry and I wasn't even remembering things that happened. My mouth was incredibly dry. I never connected all these symptoms. I thought my eyes were just getting worse and that I must be so thirsty because of my allergies."

—Ryan Kasmier (San Francisco, California)

How do you find humor in diabetes?
"I think it's the natural comedian in me. I've always been able to find the absurd in things even if it's serious. And diabetes has a ton of absurdity in it. I was diagnosed when I was twenty-five and the things I've been through between doctor visits, eye surgery, and prescriptions have given me tons of premises.

"Simple answer is, what else can I do? I've been depressed and suicidal because of what I have to go through with this condition and I'm still here. I know I can't tell every person with diabetes to 'be like me' because I don't have all the answers. But it can be done. So like I said before, you have to stop and look for the absurd. Once you get off that path of gloom, you can laugh. You just have to allow yourself too. And finding a community helps immensely. That had to be the biggest thing that helped me. I was diagnosed in 1989 and there was no online community for much of anything back then. So I was quite isolated. Thank goodness for Windows 3.1!! Or was it '95?"

—Chelcie Rice (Atlanta, Georgia)

"When I first was diagnosed, I had just started doing stand-up at open mics. I had some open mic level jokes about it:

"'Two things you learn when you get diabetes. You have to eat right, and Mrs. Fields is the Antichrist.'

"'I have to take shots now so I give myself a little prick every day. My wife is still resisting it, except on my birthday.'

"So humor, as bad as it was, was definitely a way to deal with it. When I'm on the road I can't eat like a regular 'road' comic so I get teased by the other comedians. It's always good natured and fun."

—Dave Mencarelli (Reno, Nevada)

"I'm able to laugh because it's surprising the amount of people that don't understand how diabetes works. And everybody tells me how to handle my diabetes or how to get rid of it when they don't have it. It's an incurable disease and I'm a type 1 diabetic. I didn't eat poorly, my pancreas just decided it didn't want to produce insulin anymore."

—Geoff Grant (Las Vegas, Nevada)

"Well I personally had always been good at finding humor in a dark place. And I was self-aware enough to know how much of this was my own doing. So I saw the humor in my own shortcomings."

—Greg Asdourian (San Francisco, California)

"Type 2 is a lifestyle disease, so I do a lot of self-effacing jokes about getting it. I'll say on stage, 'I don't mean to brag because you can already tell. But I do have type 2 diabetes. Yeah, the one you had to earn.' It's a disease you give to yourself, so if you can laugh about it, it's like you're forgiving yourself a little."

—Jason Ward (Boise, Idaho)

"I try to find humor in as many things as possible. Like anything, it's really people's reactions that I find a lot of humor in. I am not overweight, some would say thin, and so when I disclose to people that I have diabetes, I overwhelmingly get this response, 'How? You're so skinny.' That really ignores a lot factual information. First of all, I could be type 1, in which case many, if not most, are thin. Second, even though I don't have type 1, I have been told by many medical professionals who have looked at my blood and liver, etc. . . . that my diabetes is purely genetic. It was essentially unavoidable. My mom and brother both have it.

"Often, people also try to be 'helpful' once you've disclosed it. I was at a birthday party and the host was passing out cake and totally skipped over me saying, 'Oh, Ryan can't have that, he has diabetes.' I wanted to respond by saying, 'Bitch! There are plenty of people who should be skipping the cake but I'm not one of them.' It's such a one-dimensional view of diabetes and it's quite ignorant. If I know there's going to be cake at party, I will eat a hard-boiled egg and walk for an hour so I can indulge, but I'm having the damn cake."

—Ryan Kasmier (San Francisco, California)

"You find humor in this condition by looking at alternatives: [losing] limbs, comas, and blindness. Make fun of what scares you. Others can relate to your honesty. I try like hell but sugar is so addictive. Not death addictive. I wanna grow old healthy."

—Zorba Hughes (San Francisco, California)

What advice would you give to someone diagnosed with diabetes?

"I would tell them to grill their doctor constantly. Make sure he or she is making an effort for you to be well. Your doctors work for YOU. If they don't perform to your liking, fire them. Educate yourself constantly. Even if it seems ridiculous, because the truth is there's a ton of false info out there and we need to know how to weed it out. Lastly, get online and find a network/community to join. Never isolate yourself because there's a ton of people that can help you even if it's just a kind word of encouragement."

— Chelcie Rice (Atlanta, Georgia)

"My advice is that you can still eat whatever you want but the hard part is keeping disciplined. It's tough when you are diagnosed later [in] life because you are used to eating whatever you want and how often you want. But think of this, coping with not being able to eat jelly beans as often as you like is much easier than coping with losing a leg or your eyesight. And if you do eat sugar-free candy, watch it, too much gives you diarrhea. I found this out the hard way, you're welcome in advance!"

— Geoff Grant (Las Vegas, Nevada)

"For lack of a better term, have a sponsor or someone in your corner. My wife has been on top of me the whole time about helping me eat right and stay accountable of my health."

— Greg Asdourian (San Francisco, California)

"It's okay to live with diabetes. It's doable. But you have to make some changes to your lifestyle if it's lifestyle caused. I haven't had a soda or sweet tea in a long time, I've given up candy, and I'm in the process of cutting out all added sugar and processed foods in my life. And my fasting glucose levels have been the lowest they've been in four years."

— Jason Ward (Boise, Idaho)

"Take your sugars (I just like saying like a Tennessee Williams character, but whatever makes you take it) and eat accordingly. I actually don't find it too hard to manage and I'm told by my doctor that I'm doing great. But I totally understand how hard it can be for some people. I'm pretty lucky in that once I got everything under control and got on the proper dosage of medication, my hunger tends to match up with blood sugar. I know it's not that way for everyone. Also, you have to move. You don't have to work out hard but you have to move and walk. It makes such a huge difference in your blood sugar readings."

—Ryan Kasmier (San Francisco, California)

"When I was diagnosed with diabetes I was a little upset. I avoided all the negative things and now this. Sort of like finding out you're related to Dick Cheney. But I learned to cope with it, that's what I'd say to all those who inherit this condition: one day at a time."

—Zorba Hughes (San Francisco, California)

Got any diabetes jokes you can share with readers?
"I never got bullied the same as other kids—my house never got egged. It got cupcaked."

—Geoff Grant (Las Vegas, Nevada)

"My doctor very solemnly sat me down (which is never a thing you want a doctor to do by the way) and said, 'Greg, I don't know how to tell you this. But you have type 2 diabetes.' I looked up at her and said, 'Wow, you must be a detective. And I don't think I want to keep paying to see a doctor that is this surprised by me having diabetes."

—Greg Asdourian (San Francisco, California)

"I got diabetes the same time I got asthma when I inhaled a bag of powdered sugar."

—Jason Ward (Boise, Idaho)

"I would get into some experimental trials for diabetes, but I'm just as afraid of sugar pills as I am of experimental medication."
—Jason Ward (Boise, Idaho)

"When the doctor said I have prediabetes I thought doesn't everyone have prediabetes? Said I had to start putting more green, red, and yellow foods on my plate. Now I'm addicted to Skittles."
—Zorba Hughes (San Francisco, California)

OKAY, SO WHAT IS SO IMPORTANT ABOUT, UM . . . HEALTH?

HEALTH IS ONE OF THOSE SUBJECTS, LIKE PLEASURE, WHERE I feel the importance is inherently obvious. I am pretty sure that I do not have to convince anyone that health is important. There are reasons we have phrases like "at least I have my health." Are you enjoying all that life you are currently living? If so, be sure and thank your health.

I think it is important for us to realize that laughter is a healthy human activity. Sometimes we fall into the mistake of taking life too seriously and lose sight of this. Many people consider playful behavior and humor as being childish and abandon these as they get older, calling their newly adopted seriousness "maturity." This is a misconception, a belief that is causing harm to a lot of people. As we age, there is nothing that says we have to be mature. All we really have to do is pay our bills. Responsibility and maturity are not the same thing, yet many people either confuse the two or believe they should go hand in hand. I am very responsible and accomplished, yet if you spend any time with me you will probably witness some immature behavior. Hell, even in writing THIS BOOK I told a fart joke. Clearly, it would seem to some that I've got some "growing up" to do. I don't know if my grandmother is one of those people, but I do know that she did not approve of my grandfather's cartoon habit. And why would she? Cartoons are for children, right? No. Funny is funny at any age. Our tastes change throughout our lives, but if something makes you laugh, then enjoy it. Laughter is healthy.

I enjoy some cartoons too, although I do not make a point to watch them every day like my grandfather did. My favorite show is *Family Guy*, a modern adult-oriented cartoon (although now that I think about it, if you needed that description you probably are not reading this book). Anyway, one day when I was teaching a course at a college in Ohio, I had a discussion about *Family Guy* with a few of the students before class. After the lecture, another student approached me and said, "I was surprised to hear that you watch that show, you seemed so deep and intellectual." Her comment caught me off guard for a couple reasons. First, there is a lot of intelligent satire in that show and some great criticisms of pop culture, so clearly she was missing a lot if she's seen it. But more importantly, behind her comment was a belief that intelligent, educated professor types should not watch cartoons. As if instead, I should be spending my free time at the opera, conjugating Latin verbs, or figuring out math theorems. It sounds ridiculous, but many people believe that as they grow up, as they achieve more, life should be increasingly more serious.

A therapist once consulted with me about a patient he was treating for depression. He had been working with her for a while without any noticeable progress and he asked me about humor therapy. I'll talk about humor therapy toward the end of the book, but for now, just understand that it sometimes involves "prescribing" a patient to enjoy a daily dose of comedy. He wanted my advice on what type of comedy he should prescribe. When I make recommendations like this, I want to get to know the person's existing interests so that I can help match them up with something they are more likely to enjoy.

I asked her, "What kinds of movies do you watch?" and she said she did not go to the movies. In fact, she hasn't been in years. So then I asked, "What are some of your favorite TV shows?"

She told me that she did not watch much TV. The only exception to this was the news—when she turned on the TV, it was usually for the news. Once in a while she would watch a documentary on Netflix. I started to get an interesting picture of her habits, so I probed a little deeper. "What kind of music do you enjoy?" I asked. She said she did not listen to music very often, but would sometimes put on some classical music when she was reading. "What types of books do you read?" I asked, and she replied that she almost exclusively read nonfiction, and mostly biographies. So here was a person who does not watch any movies or TV for entertainment and barely listens to music except for when it provides ambient noise to help her study. I told her that it seemed to me she was deliberately excluding a lot of things from her life that are known to make other people happy. She agreed and when I asked her why, she replied, "I think it is beneath me."

This patient was very educated and very accomplished in her chosen field. She felt that popular entertainment like movies, TV, and music were a waste of her time. If she was going to read or watch a program, it had to be deemed educational or otherwise enriching. And she wondered why she suffered from depression.

By the way, for that patient I recommended that she ease her way into comedy by watching *The Daily Show* on Comedy Central (or online). I explained that it was a satirical news show that covers real events but in a very humorous way. Because she was already a news watcher, this would not seem like too much of a departure from her regular viewing habits and maybe it would be enriching enough that she could connect with it.

Please do not misunderstand what I am getting at here. I am not telling you to start watching stupid movies or TV shows, or to start reading humor magazines. What I am telling you is to recognize the value of humor and laughter. If something does not make you laugh, then don't waste your time. But do give yourself the freedom to explore options. Remember that laughter is healthy.

Another thing I would like to point out is that the health benefits of laughter that I previously listed are not unique to laughter. I would love to say otherwise and therefore increase comedy club attendance as soon as this book hits stores, but that would be a dubious claim. There are many behaviors that human beings regularly engage in that boost our immune system, help lower our blood pressure and blood sugar, and work out our respiratory system. Probably the most important of them is EXERCISE. Perhaps you have heard of it? Exercise is extremely healthy.

And speaking of that . . .

LAUGHTER IS EXERCISE

Yes, LAUGHTER IS EXERCISE. NOW BEFORE YOU GO CANCELING your gym membership keep in mind two things that I have previously shared with you: 1) I laugh a lot; and 2) although I am actively working on it, at the time of this writing, I am pretty overweight.

Laughter is not strenuous exercise, although it can be. You are not going to burn off your spare tire through laughter. It isn't going to shrink the mud flap or get rid of your muffin top, and certainly is not going to replace other forms of physical activity. However, it does offer a significant workout for specific muscle groups, making this another benefit worth noting.

Recall that earlier when I discussed laughter I described it as our body's response to humor. The nature of that response can vary from a simple smile and chuckle to full-on arm waving and leg stomping, rolling on the floor. All that activity? That is what we call exercise.

A low-intensity laughter uses at minimum fifteen different facial muscles. Not to mention our respiratory muscles and our diaphragm. A more extreme ROFL is going to engage all of those, plus our arm, leg, and back muscles. Laughter can be a serious workout, if we do it long enough.

Usually periods of laughter do not last very long, but occasionally we find something so incredibly hilarious that we might bust a gut or split our side. Both of those phrases come from the idea that we are capable of laughing so hard that we will literally experience pain in our abdomen. That pain is from overworked muscles, muscles that generally are not used to that level of activity. Laughing can be very strenuous.

If you have ever been to a professional comedy club for a show, chances are that show was somewhere near ninety minutes long. After years of trial and error, the comedy club industry has settled onto that as the magical show duration. Rarely, comedy shows will go for a full two hours, but they will hardly ever go longer than that. This is for professional clubs; some miserable open mic nights will go for three to four excruciatingly painful hours. These "comedy death marches" aren't much fun for anyone, and by the end of the night the only people left in the audience are the last few comedians waiting to perform. The reason for this is that the audience literally gets tired of laughing. An hour-and-a-half worth of strenuous laughter is about all that most people can handle before their bodies get tired.

Comedic movies also clock in around ninety minutes. Comedies tend to be among the shortest movies Hollywood spits out, whereas the duration of dramas and other genres seems to keep increasing every year. Audiences will sit through twelve hours of Hobbits walking through New Zealand (seriously, forget the magic ring, those movies are about walking), but have a shorter tolerance for comedies. On TV, most comedies are half-hour shows. How many comedies go for a full hour? How many dramas are wrapped up in thirty minutes? If you answered "not many" to both of those questions then you were reading my mind.

More and more people are recognizing the value of laughter as exercise. For the past twenty years, thousands of people across the globe have been practicing Dr. Kataria's laughter yoga as a form of physical activity. It may not fully replace our other workouts, but it isn't trivial either.

LAUGHTER MAKES US FRIENDS
(AND MORE)

THIS AFTERNOON MY GIRLFRIEND, SARAH, SAID TO ME, "YOU ARE so funny. I love you."

I replied, "So do you love me because I am funny?"

"Yes, among other wonderful things. Many, many other things . . . you are so great, you seriously rock my world." (Admittedly, she may remember this line differently, but hell, it is my book.)

"I think I want to use this quote in my book, can you say that you love me because I make you laugh?" I asked.

"Okay, if it is for the book. Yes, I love you because you make me laugh."

"Thanks," I said.

"And many other things. You are way better looking than George Clooney," she added.

Okay, this conversation may or may not have happened exactly like that, but the point should be clear: laughter influences love. It does. In fact, it influences all sorts of human relationships, from making friends to creating teams of coworkers, As comedian Victor Borge put it, "Laughter is the shortest distance between two people."

In previous chapters I discussed how Laughter stimulates the release of endogenous opioids—feel-good, painkilling chemicals we call endorphins. And these are not the only substances produced when we laugh. I've been keeping the discussion on the brain really simple, but there is actually a whole bunch of stuff happening in our head when we respond to humor. One of the other chemicals released is a hormone called oxytocin. Oxytocin is sometimes nicknamed the "empathy hormone" or "trust hormone," and as those names would imply, it is involved in social bonding. It is released whenever we hold hands with another person, share a hug or a kiss, or get naked and engage in a little horizontal tango (yes, I am that immature). Handshakes, hugs, and humps: the "three H's" of human bonding all stimulate the release of oxytocin. It is released in women during childbirth and after with nipple stimulation (presumably by breast feeding, who is the immature one now?). Basically, it is released whenever we make positive bodily contact with another human being, so being punched in the stomach doesn't count. It is released with positive bodily contact AND through laughter. I find that intriguing.

Oxytocin is a hormone that helps us create bonds with other people and laughter stimulates its release. Laughter therefore facilitates human bonding, and there is a great deal of social benefits to laughter. I have already discussed how laughter is inherently social and that we are thirty times more likely to laugh in the presence of other people. Well, it also has a social function. Every type of human relationship, whether it is between family members, coworkers, friends, lovers, or Craigslist hook-ups, is strengthened through laughter.

Let's take coworkers, for example. Throughout my career, I have had a ton of crappy jobs. In college I flipped hamburgers, stuffed tacos, and tossed pizzas (sometimes out of a moving car)—all for minimum wage. One year I had an overnight job for which I donned a rain suit and sprayed down a butcher shop, machinery and all, with high-pressure hot water by myself. The work was terrible, and the rain suit was so ineffective that I would leave each night soaked with water and covered in tiny gross animal parts. For a summer, I suffered the symptoms of my grass allergy and massive amounts of fire ant bites to lay sod for a landscaper. I've also sat in cubicles for ten hours a day or more, illuminated only by flickering fluorescent life-sucking lights, turning computer keys into carpal tunnel syndrome. After working all of those low-wage, manual labor positions though college, I would say that the crappiest jobs I have ever had came after graduation.

There are a lot of variables that influence how satisfied we are with a job. The nature of the work itself is a factor; some jobs are objectively more appealing than others (I would imagine flipping condos is more exciting than flipping hamburgers, although my financial advisor wonders about the debt to equity ratio of the condos). Our own personality disposition plays into it; some jobs are a better fit for us than others (as a "creative" I am not a good candidate for the mundane or overly structured). Compensation is definitely an issue; more pay seems to make any job more attractive to a point (although some of my least favorite jobs have had the biggest paychecks). Also, how happy we are influences our job satisfaction. Conversely, how satisfied we are with our job influences how happy we are.

When I look back at my resume, I realize that for me, one of the biggest factors in my satisfaction was whether or not I developed good relationships with other people on the job. The work, the compensation, or the benefits—none of that seemed to influence my happiness as much as my coworkers. For example, I mentioned I used to flip burgers when I was in college. I did so at two of the most popular fast food chains and it doesn't take a genius to figure out which two. My main duties were to work the grill stations but as part of a team I was trained to do it all. At one burger joint, my team was a blast to work with. We got to know each other very well, joked around a lot when things were slow, and generally had a great time in the kitchen. In fact, some of those friendships lasted for years after moving on. To this day I still look back with fondness at my time slipping on french fry grease in my polyester uniform smelling like burgers. At the other place, things were more serious and I had several coworkers that I never interacted with. I quit that job without ever looking back. Clearly one of those jobs made me happier, the one that gave me more free hamburgers. Kidding, it was the one with the friends.

Ultimately flipping burgers among friends is not where I hoped my life would end up. I eventually graduated and entered the workforce with some abbreviations behind my name. Early in my career, I had a few jobs that were pretty similar. They had the same type of work, same type of cubicle-farm office environment, and similar compensation. However at one company the office was full of energetic, positive people who spent time on their breaks chatting between cubicles and getting to know one another. Laugher was common, and I made a lot of close friends in that office. Some of them I still count among my best friends, despite the fact that we all moved on a long time ago. In another office, I worked with a group of coworkers who never shared anything personal, never got to know each other, and never laughed. In fact, I had an officemate who refused every one of my attempts at establishing a relationship. For a year, I literally shared an office with a guy so humorless and so impersonal that I don't even remember his name. I wish I could say his behavior was unique, but it was consistent with the overall corporate culture of the sucky office. Guess which of those jobs I liked more?

For me, laughter has been crucial in my liking my workplace. I know others would say the same. People who laugh together are more likely to develop close bonds. People who develop close bonds are more likely to cooperate with one another and be interested in things like teamwork. Laughter turns coworkers into friends, and having friends at work is a key factor in job satisfaction. Looking back, if I did not have the financial obligations, I would much rather have flipped burgers with my friends than sat in an office with what's-his-face. Wow, I can't even picture what he looked like anymore, and I spent a year in his immediate vicinity.

There is a seriousness with which some people approach their work that prevents them from establishing good relationships at the job. Granted, if we like getting paid we need to do what is expected of us, but that does not always mean we have to be humorless in the workplace. My girlfriend is a therapist, and laughter is a regular part of her work as it helps establish trust with her clients. She recalls one day when after working with a client and laughing, a nurse said to her, "You guys were having entirely too much fun in there. You must not be working that hard." Comments like this bother me because they are simply not true, and they reveal an implicit belief that all work has to be serious. That belief, I believe, can have a seriously negative impact on our corporate culture.

When I was deep in the trenches of corporate America, I used to say that teams that drink together work well together. I worked as a consultant and spent a lot of time visiting and learning about other organizations. I was able to see how a lot of companies do business and either observed, or was a part of, a wide variety of corporate cultures. One of the things that struck me was that some offices tended to have a very tense, uptight vibe. The employees seemed moderately stressed, did not relate to each other on a personal level, and almost never socialized after hours. Other offices were vibrant and full of energy; their employees seemed to genuinely like each other and happy hours were a regular thing. I mistakenly assumed that sharing a drink made the difference, but it was more likely the attitude and laughter that went with it. Teams that laugh together work well together.

Although, one of the happiest work environments I ever observed was at a brewery in Milwaukee. They had taps serving their signature products in every meeting room; they literally had BEER freely available to employees and guests. During meetings, it was not uncommon for someone to get up and serve themselves a cold one. Because I was a consultant from an outside organization, I had to be on my best behavior, but as soon as the meeting was over and the clients were gone, I was hitting those taps. Best job ever.

I did not go into all of this just to provide you with an example of how laughter makes work tolerable by creating friends. Of course that is the point, but I wanted to go into this discussion a little more deeply because I know that some of you may work in unfriendly environments or have experiences similar to those I described. We spend a good chunk of our lives at work, and it might as well be a source of happiness for us. However, as I've previously mentioned, for a lot of us it is our largest source of stress. According to the Gallup organization, only 30 percent of American workers are engaged on the job. This means that there are a whole lot of us that hate our crappy effin' jobs.

By the way, I have also had some amazing jobs. Being a professor is pretty great, and so is being a comedian. They are a lot alike except that as a comedian there are fewer people taking notes. They really are similar occupations, drawing upon skills of communication and presentation. In fact, the first time I performed stand-up I was surprised by how much it felt like teaching a class. But as much as I enjoy my current occupation, I also really enjoyed making tacos and pizza.

By releasing oxytocin, laughter helps us bond and establish trusting relationships with other people. It turns strangers into friends, and strengthens relationships among coworkers. It promotes team building and group bonding. And it does so much more.

In the 1988 movie *Who Framed Roger Rabbit*, detective Eddie Valiant (Bob Hoskins) asked the beautiful Jessica Rabbit (voiced by Kathleen Turner) what she saw in her goofy husband and she replied, "He makes me laugh." That's right, laughter can be a major factor in physical attraction. Once someone makes us laugh, we find them more physically attractive than we would have before. If we make someone laugh, we become instantly more attractive in their eyes. My girlfriend is ridiculously beautiful and although I am not ugly by any means, let's just say that I am extremely lucky that she thinks I am funny.

Laughter facilitates physical attraction and oxytocin helps establish bonding relationships. Therefore laughter helps us fall in love. Obviously, not everyone laughs so not every relationship is going to include laughter, but it helps. It helps a lot. Comedian Yakov Smirnoff, who I mentioned earlier, is often quoted as saying, "If love is the treasure, laughter is the key."

In general, men and women both find a sense of humor attractive when evaluating a potential partner, but there is a very interesting sex difference that I would like to discuss.

First, let me just state that it is a widely held belief that men have a stronger sense of humor than women. Now, I know I run the risk of losing all my female readers after that last statement, so let me be clear that I do not share this belief, and research does not support it, but it is there. It is easy to understand how some people may come to this conclusion with a casual glance around the world of entertainment. Most comedy movies are male-oriented and so are most comedic TV shows. There are some fantastic female performers out there, such as Tina Fey, Ellen DeGeneres, and Wanda Sykes, and as I write this Amy Schumer is one of the hottest stars in Hollywood. However, most professional comedians are men. A lot of female comedians are very vocal about their perception of sexism in the industry, and a lot of male comedians are sexist. Wanda Sykes even mentioned that she thinks "it's harder for women because comedy is so opposite of being ladylike." Whatever the reason, professional comedy is overwhelmingly male dominated, and most demographic estimates hover around 80 percent male. The amateur ranks are also mostly male, and you can even think back to when you were in school. I bet your class clown was some dumb guy. If you were the class clown, you are probably a dude. Given all of this, it is reasonable to see how some (not me!) may think men have the better sense of humor.

On a side note, although the comedy industry is dominated by men, some of my favorite comedians are women. I mentioned Wanda Sykes. By the time I first saw her perform live, doing a quick drop-in set at the Comedy Cellar in New York, I had already been a fan of hers from seeing her on TV. She is one of the smartest and most interesting voices in comedy, period. She can command a stage so well; I once watched her follow Stevie Wonder at a party for outgoing Los Angeles mayor Antonio Villaraigosa. Following musicians is extremely difficult for a comedian, and following the likes of Stevie Wonder is damn near impossible. Wanda pulled it off, like the amazing professional she is.

Actually, when it comes to sense of humor between the sexes, it depends on how that humor is defined. Men and women in general find different things funny. I've mentioned the slapstick physical comedy of *The Three Stooges* a few times already. Men tend to enjoy this form of humor more than women do. My grandfather was a good example. He would laugh his ass off watching *Tom and Jerry* cartoons while my grandmother would be in the other room not quite seeing the humor in it. I imagine that most of the core audience for the TV show *Jackass* is made up of men as well (although I have an ex-girlfriend who loved that show, mainly because she liked seeing men in pain).

Men and women laugh at different things, but here is the first piece of the sex difference puzzle that I would like to present to you: in conversation, women laugh more often than men. Conversely, men are more likely to make people laugh. This was demonstrated quite well in an interesting piece of research by Robert Provine (2014). Using naturalistic observation techniques, in other words literally treating human beings as animals to be observed in their natural habitat (in this particular case the natural habitat was a coffee shop), the researchers observed the conversational behavior of people in groups of two. Without knowing anything about the nature of the relationship of the people under observation, they sat at a nearby table and took note of how often each person spoke and how often each person laughed. They found that when a woman is speaking to another woman, she is laughing 86 percent of the time but her listener is only laughing 50 percent of the time. When a woman is speaking to a man, she is laughing 88 percent of the time, while he is laughing only 39 percent of the time. On the other hand, when a man is speaking to a woman, he is laughing 66 percent of the time but she is laughing 71 percent of the time. When a man is talking to another man, the speaker is laughing 76 percent of the time, the listener 60 percent of the time. Those eight statistics reveal the two interesting patterns I mentioned above: regardless of whether they are speaking or being spoken to, women laugh more often than men. Also, regardless of who is listening to them, men are more likely to elicit laughter in their audience than women.

I am very fond of simplification, something that you should already know because you've made it to this chapter. I like to combine the above two findings into one very simple and general statement: men make women laugh. Remember, there are many, many, many exceptions to this rule, but in general, ladies, we are nothing more than goofy fools for your amusement. I am sure that we may have other uses as well, like peeing out campfires or reaching the top shelf, but with regards to communication men make women laugh. It's as simple as that.

I find that fascinating, and it is only the first piece of the puzzle.

It is important to note that I AM simplifying quite a bit here and that not all the laughter recorded in speech is humor-related. To keep me in check, my friend and New York comedian Ariel Elias offered the following counterpoint:

> Women may laugh more than men for a number of reasons, but it's important to keep in mind how many of our behaviors are learned. Throughout our lives, we're taught over and over via TV, movies, and our own mothers that if we want to attract a man or let him know

we're interested, touch his arm and laugh at his joke, even if it's not funny. For a long time, a woman's role was to stroke her man's ego, and although this attitude seems antiquated, I think laughing a lot is a result of that legacy. You're not actually that funny. Remember, we're told to smile while we're walking down the street, to seem pleasant and approachable at all times. The term for a woman's face that isn't smiling or laughing is literally called "resting bitch face." I have never in my life heard a man's face described the same, or even "resting asshole face," which is too bad, because it's pretty accurate for some of you.

Before I continue, I would like to state for the benefit of my readers that are unfamiliar with scientific practices, observational research like that is conducted by social scientists all the time. It might sound a little creepy, to think that some scientist at the table next to you is keeping tabs on your private conversation; however, it is a valid and in some cases completely necessary research method. We can't reliably ask people to tell us how often they laugh, and studying them in experimental settings would probably not yield generalizable results. Scientists working in the field are careful to make their presence unknown to those being observed (this is important because it would severely taint the results) and only record data important to their research objective. Provine was not recording the topics of conversation, just timing the laughter.

So men make women laugh, but why? Well, I have given you a few examples already with the conversation I had with Sarah and the Jessica Rabbit quote. The second piece of the puzzle is that women, in general, like men that make them laugh. I think actress Audrey Hepburn articulated this well when she said, "I love people who make me laugh. I honestly think it's the thing I like most, to laugh. It cures a multitude of ills. It's probably the most important thing in a person."

It is important to understand that when it comes to evaluating the traits that attract us to a potential partner, both men and women value a sense of humor, however women place a greater emphasis on a sense of humor than men do. There are several large-scale cross-cultural studies that have surveyed mating preferences and they all confirm that women in general desire a sense of humor in men more than men desire the trait in women. Let's examine data collected as part of a large-scale international survey conducted by the British Broadcasting Corporation and analyzed by psychologist Richard Lippa (2007). Utilizing data from more than one hundred thousand men and ninety thousand women across fifty-three countries worldwide, they found that the number one attribute that women look for in potential mates is a sense of humor. It is at the top of the list, followed by intelligence, then honesty, kindness, values, communication, dependability, good looks, and facial attractiveness. I like to joke that it is a good thing women put such a low priority on physical attractiveness or I would never score a date. My girlfriend insists that she thinks I am cute, but we all know she loves me because I make her laugh. Cross culturally, women like men who make them laugh. Other studies show very consistent results Men also like women who make them laugh, but humor is less of a priority for men. According to Lippa, the top trait that men say they are interested in is intelligence. This is followed, not surprisingly, by good looks (actually I was surprised this was not first) and then humor. Men seek a sense of humor too, but put less of an emphasis on it than women do.

It is important to understand that these preferences are observed cross culturally. As I mentioned previously, recognizing that a phenomenon occurs across cultures supports the idea that the preference may be innate and therefore potentially biological.

I have been aware of this sex difference since my days in graduate school, and it has always bugged me as I had never come across what I felt was a satisfying explanation for this difference. The differential emphasis placed on good looks (men prioritize them but women view them as far less important), for example, has been explained quite well by evolutionary psychologists as it relates to reproductive fitness. I am not going to go into that here, but I at least want you to be aware that there are explanations. Throughout my education, I had never encountered a good reason why women should value humor in a partner more than men. In fact, when I posed this question of why women want a man who makes them laugh to one of my lab partners he suggested, "Well, who wants to be with a boring dude?" (He was gay by the way, so he might have a little insight.)

One thing we do know from the battle of the sexes is that if one side expresses an interest in some trait, members of the other side better start having that trait if they hope to get any action. Think of peacocks. For some reason, peahens love long-ass colorful tail feathers, so peacocks with exaggerated tail feathers to shake were able to score more partners and father more offspring than those shorter-tailed losers. Now we have a bunch of males with stupid (but beautiful) feathers making it difficult for them to fly. The same sexual selective pressure used to develop certain physical traits over history also applies to behavioral traits. What happened to the peacock's ass happened to the human male brain.

So to put the puzzle pieces together, I can simplify even further and state that in general, the reason men make women laugh is because women find humor attractive. That may be a valid statement and keep in mind, I am discussing generalizations, but it still does not answer the question that has bugged me since before I became a doctor. Why is it that women prefer men that make them laugh?

Here is another piece of the puzzle, and this piece was missing when I was in graduate school. As I have previously discussed, laughter is pleasurable. What was that, you missed it? The chapter was literally called "Laughter Feels Good." To recap, it feels good to laugh. Laughter increases activity in the nucleus accumbens, an area in our brain that we often refer to as our pleasure center. When we laugh, that activity is translated by our brain into the subjective experience of pleasure. Now here is the kicker, this activity is more pronounced in women than in men.[81] That implies that the pleasure experienced by women as a result of laughter is greater than that experienced by men. Both men and women experience pleasure when they laugh, but women experience more. If you have not yet picked up, I am a man. I love to laugh and I laugh a lot. When I laugh it feels good, but when women laugh it must feel AMAZING. Unfortunately I will never know what it is like to laugh as a woman. Life can be so unfair; I'll have to be content with being able to pee out campfires. Maybe Caitlyn Jenner will be able to find out what it's like.

Making someone laugh is extremely rewarding as well, as any comedian can tell you. I am unaware of any research that shows this, but it would not surprise me at all if men showed a greater increase in activity in the nucleus accumbens than women when making others laugh. It would certainly make sense, given everything I have discussed so far, if men were to receive a greater reward in the brain from being funny than women. The problem I see with conducting this research is imaging a person's brain while that person is making someone else laugh. I can't imagine sticking someone in an MRI tube and then telling them to be funny. And to what audience, the researchers? How rewarding would it be to make a couple of lab nerds laugh? If you can figure out how to design such an experiment, you are welcome for the dissertation topic. Just make sure you cite me.

So women get this enormous reward benefit when they laugh. It now makes a whole lot of sense why they show a preference for men who make them laugh. Let's use a food analogy for a moment. In my brain, chocolate stimulates way more activity in the nucleus accumbens than kale, the most disgusting of the vegetables. Throughout my lifetime of learning the relative reward values associated with chocolate and kale, I understandably have developed a preference for chocolate when given a choice. Now, if kale had a way to modify itself to be more like chocolate . . . okay let's not get lost in the analogy. Besides, even chocolate-covered kale tastes terrible.

Men make women laugh because women prefer men with a sense of humor because laughter is much more rewarding for women. How is that for an exercise in simplification? As long as women enjoy laughing, men will continue to cultivate the skills to make them do so. This reminds me of a quote from author and Facebook personality Charles Orlando: "The quality of a man isn't measured by the size of his bank account, but rather by the size of the smile of the woman on his arm." This is a nice sentiment, unless of course it is the size of his bank account that is making her smile.

A sense of humor is attractive, and the act of laughing produces oxytocin, which helps us strengthen our bonds. Therefore laughter does not just open doors for us, it helps keep us in the room. Yakov Smirnoff has an interesting perspective on the importance of laughter, not just for developing a new relationship but for sustaining it as well. For more from him, I'll refer you to my later chapter titled "Laughter Helps Sustain Relationships: My Chat with Yakov."

Again, this is all generalized. There are exceptions to everything. There are men who do not care about their partner's looks, and women who are all about them. There are lots of boring, unfunny men in the world, and lots of great female comedians. Sarah also makes me laugh a lot.

NOW THAT THE QUESTION HAS BEEN (SORT OF) ANSWERED . . .

I MENTIONED THAT THE QUESTION OF WHY WOMEN PLACE MORE value on their partner's sense of humor than men has been bugging me since graduate school. Well, now we have an answer in the form of their greater reward response to laughter. Laughter feels better for women.

Unfortunately, as is the case with science, whenever we come up with an answer all it does is generate more questions. So now I wonder, why would the female brain reward women for laughing more than the male brain? Obviously this seems to guide their partner preferences, but why? What do women have to gain by being attracted to, and therefore becoming more attached to, men who make them laugh?

There are a few theories. Some have suggested that a sense of humor is a trait that is highly correlated with some other desirable trait, perhaps one that is not otherwise observed. It is possible then that by being attracted to men who demonstrate a good sense of humor, women are increasing their likelihood of getting that other trait. This sounds reasonable, but I don't buy it. Others, like my lab partner in graduate school, have suggested that the entertainment value of a man with a sense of humor provides boredom relief during periods of low stimulation. That's great and probably true, but there is no reason for this to be a one-sided benefit. Women can entertain their men too.

I believe that the reason both men and women find a sense of humor attractive is because of everything that I have previously discussed. Laughter makes us happy, feels good, decreases pain, reduces stress, and makes us healthier. All of those benefits are bestowed upon us if we are motivated to find a partner who makes us laugh. If we are attracted to people who make us laugh, then those people are in turn going to make us healthier. This makes sense for both sexes, which is why humor is a top trait for men as well as women. However, women express a desire for this trait more than men do.

I mentioned that these preferences are cross-cultural and that whenever we identify that a phenomenon occurs across cultures, it increases the likelihood that it might be something innate or biological. To paraphrase the old debate, it is more likely nature than nurture. Well, if that is the case, then there must be a biological reason underlying our desire for a sense of humor—and there is: increased physical health. The fact that woman show this desire more than men could imply that they value the health benefits their partner gives them more than men do. Or it could imply something more important. Women are more often the caregivers of children, and it may be children that have led to the evolution of this difference in our brains. By seeking a partner who makes them laugh, women are not only ensuring that they will potentially be healthier as a result, but that their children will benefit as well.

This is all speculation, but I think an explanation based in reproductive responsibilities makes a lot of sense. You are welcome again for your dissertation topic.

WHAT IS SO IMPORTANT
ABOUT MAKING FRIENDS?

I WILL ADMIT, I HAD A REALLY TOUGH TIME EVEN TYPING THAT question. It seems like such an obvious answer that I almost did not include it, and then I realized that a lot of people out there could use a reminder. Friends are extremely important. We need them to keep us out of trouble, serve as our wingman, hold our hair when we are praying at the porcelain throne, and occasionally take one for the team. When I tour the country speaking about happiness, I am very fond of telling my audiences that outside of our own heads, other people represent the biggest potential source of happiness we have. Now that is not to say that all other people are going to make us happy; some people are going to make us damn miserable, but the presence of other people is very important to our health and well-being.

We are an inherently social species. Just take a look at your fingers for a minute and imagine using those scrawny dull digits to catch and kill prey all by yourself. Try to imagine your weak jaw and flat teeth trying to puncture the animal's hide. Now imagine those awkward two legs of yours trying to help you chase down and catch that potential prey, because neither of the first two things actually happened. Now imagine practically starving to death in the wild because lone human beings are horrible natural hunters. If you still have some imagination left, give yourself a couple of friends and suddenly it is much easier for you to imagine a full stomach. Human beings become much more effective hunters in packs, increasing our likelihood of eating and surviving.

Obviously in the modern world we are capable of fending for ourselves to an extent. I would argue that we can never truly live independently thanks to the way our economy works, but we could probably live without friends. Many people do. The problem is that our brains evolved in a world where having other people around us increased our chances for survival, so any behaviors that facilitated this and their underlying biological mechanisms were passed down to us. Our brains are hardwired for social activity; we have whole regions of brain devoted to language acquisition and usage. What value does language have if we can live just fine by ourselves? That is a big chunk of our brain to sit there unused. We also experience a great deal of activity in our reward center when we engage in social behavior, making social interactions a source of pleasure and therefore more likely to be repeated. So yes, we might be able to sustain our life without the benefit of friends, but we are going against the nature of our brain and it's probably not going to be a very good life.

Other people make us happy
Imagine the opposite of having friends. We become socially isolated and loneliness kicks in. Loneliness is an emotion that if left unchecked can lead to the development of depression. As I have discussed, depression is a behavioral disorder and social withdrawal is one behavior that is a major contributor to its development. Chronic pain syndrome is another and you might recall the story of my uncle who had become socially isolated.

Other people help reduce our stress
Looking back on my life, I was able to successfully work through every setback and tragedy I have ever experienced thanks to having a good network of friends. Whether it was job

loss, car trouble, illness, or heartbreak, having people in my life helped make those little setbacks more manageable and a lot less stressful.

Other people encourage physical activity

I'm not just referring to sports, which clearly require the presence of other people, but any activity. Friends motivate us to leave the house, get off the couch, and, you know, do things. This in itself is important to both our physical and mental well-being.

Other people make us laugh

Remember that we are thirty times more likely to laugh at something when there are other people present. If laughter is so widely beneficial, then having friends increases the benefits.

So I just went through a very quick list of why having friends is so important to us. This is an argument that I don't think I have to work too hard to make. All aspects of life are better with friends, and their presence can influence our health. Earlier I used the example of how having friends at the job makes work a much more rewarding place to be. The same can be said for school, church, home, prison, or wherever we may find ourselves.

The fact that friends are important is something I even have to remind myself of once in a while. Sometimes I get focused on certain projects and my time is consumed by them. For example, um . . . writing this thing you are currently reading. Coincidentally, after working on the last chapter, I complained to Sarah that I hadn't really interacted with anyone in a few days and I was starting to go a bit stir-crazy. I hear a lot of writers sequester themselves when under a deadline, so maybe it is not abnormal, but I definitely felt the effects of isolation.

Humor helps us make friends. A good joke can break the ice between strangers and a good laugh can help get that socially bonding oxytocin flowing. Whenever I meet new people, almost instinctively the first thing that comes out of my mouth is a joke. It is the fastest way to establish a relationship. We are not always interested in creating new long-term relationships with people, but even short-term relationships are made easier through humor. This is the reason that public speakers often begin their speeches with a joke.

As for more romantic relationships, take everything I've said, multiply it by some extreme factor, and add sex. Speaking of which . . .

LAUGHTER IS FOREPLAY

AH SEX, THAT WONDERFUL FORM OF PHYSICAL EXERCISE THAT fills our brain with endorphins and oxytocin while simultaneously ruining our sheets. To quote George Michael (because yes, I am that old), "Sex is natural, sex is good. Not everybody does it, but everybody should."

I promise I won't start off this chapter with another quote from my girlfriend, Sarah, although there have been countless times where she has laughed at me during sex, mostly while pointing. Okay, that might have been too much information. I promise not to get too personal, my mother is reading. I did not originally intend for this to be independent of the love chapter. However, as Sarah and I were involved in some hardcore book research one night (okay, I'm sorry, Mom), it occurred to me that there may be more to say here than I first thought.

I will share with you one more piece of information about my background: I have not always focused on humor and laughter. I was interested in a lot of things in college, and like most students, my primary interest was sex. Literally, though — as in I researched and taught about sexuality. If this were a bar, I would have a very different answer to "why I studied sex," but given that Sarah is probably reading this too, I'll give you my straight answer. When I entered graduate school for neuroscience I had no idea what I wanted to study, but I wanted to focus on behaviors with direct relevance to survival. I started working in a lab studying the effects of amygdala damage on both feeding and sexual behavior in rats, and eventually shifted my focus to another brain structure and its relevance to sexual motivation. From rat studies, I jumped to human parallels and that led my work into a lot of different directions. I studied pain sensitivity and sexual arousal, the patterns of sexual arousal, mate selection, individual differences in physical attraction, BDSM (an offshoot of my pain research), and even cybersex as this was the early days of the Internet. I found being in graduate school to be somewhat like being a fat kid in a candy store (and I am a fat kid). For the first time in my education, I could literally study anything I found interesting! And what is more interesting than sex? My own research would suggest that for some people, lots of things are. However, I'd always been a little more interested in the topic than most, and having the opportunity to pursue my interests in an academic setting was the best thing for me.

I'm telling you this because I want you to understand where I am coming from when I suggest that laughter is foreplay. A lot of people relate laughter and sexual behavior because laughter is something that both men and women find attractive (see the last chapter). Many writers have drawn analogies between laughter and foreplay. For example, some anonymous Internet dweller gave us, "The sexiest curve on your body is your smile." Poet Marge Piercy said, "Shared laughter is erotic." And there are countless other examples available through a hip new website I recently introduced to my grandmother called Google (maybe you've heard of it?). I want to draw a distinction between myself and the poets, writers, and other metaphor slingers out there. When I say, "laughter is foreplay," I literally mean that laughter may help our bodies get ready for sexual intercourse.

We are all adults here, I'm assuming. But let's do a quick inventory; there are a few basic ingredients we need if we are going to whip up a batch of booty-call stew. For the typical heterosexual version of the recipe, we'll at least need an erection, some lubrication, and a case of Gatorade (okay that is optional). Now I am unaware if anyone has ever shown that laughter contributes to any of those ingredients (although they do refer to laughter as a "social lubricant," ba-dum tsh) but there is one additional item that is not necessarily a requirement, but pretty damn important. Before I go into that, let me give you a quick summary of my research on pain sensitivity and sexual behavior.

I began by studying rats. In any group of rats, there are a small percentage of males that will not copulate with a receptive female when given the opportunity; they just don't show interest. Maybe if rats had video games and weed they would express an interest in those, but I digress. We classify these males as "noncopulators" and research has shown that their lack of interest is probably due to a lower level of sexual motivation. Interestingly, these males also have a higher threshold for pain than males that are more ready to party. Noncopulators can be encouraged to get busy by inflicting them with a little pain (don't worry, it's just a minor tail pinch with a fingernail). These two findings have lead researchers to theorize that sensitivity to sexual stimulation and painful stimulation are not distinct in the brain but rather two ways to activate a general arousal mechanism. The noncopulators were males that, for some reason, required more intense stimulation to put them over the arousal threshold in order to initiate sex.

Among us human beings, at least those of us who are experienced, I am sure that we can agree that sometimes a little pain, like a scratch, a bite, or a hair pull, can drive us crazy. In experimental research, I attempted and was able to show a parallel relationship between sexual motivation and pain sensitivity among both men and women. Using the cold pressor test I described previously, a bit of pre-tested erotica, and some survey questions, I found that the less sexually motivated a person is, the higher their pain threshold.[82] We interpreted this in the context of the animal model that perhaps males and females with a lower motivation require more intense stimulation to reach an arousal threshold. Now, could humor or laughter serve as a priming stimulus for men with lower sexual motivation? Maybe, but I am unaware of studies that show those correlations. We can all probably imagine scenarios in which laughter helps get people in the mood. Tickling makes us laugh and can be a form of foreplay, so it is highly likely that tickling a partner or being tickled helps get the juices flowing. Researchers have observed that a man's ability to control his sexual arousal response is directly related to his ability to control his laughter,[83] so perhaps the notion of laughter helping us get aroused is not completely unimaginable.

When it comes to preparing our bodies for sex, men actually have very little to do. Female bodies have comparatively more work to do before they are ready. Let's return to rats for a moment. Another point of interest is that sexual behavior can produce analgesia in female rats. It is believed that this effect helps prepare the female for the pain associated with having sex with male rats, who are not very considerate partners. The analgesic effect of sexual arousal observed in female rats has also been replicated in women,[84] suggesting that as women are exposed to sexually arousing stimuli, their bodies prepare for intercourse by becoming less sensitive to pain.

Now I am about to make a HUGE leap in logic as I attempt to connect these dots, so please hang in there. Because intercourse is a naturally painful activity, the additional ingredient I alluded to a few paragraphs ago would be a reduction of female pain sensitivity. This happens during sexual stimulation and it happens during laughter as well, arguing that perhaps, just perhaps, laughter is helping our bodies prepare for sexual activity. It might just be a literal form of foreplay. And because sexual intercourse is potentially more painful for women than for men, perhaps that also contributes to why women value a sense of humor in their partner more than men.

Of course, I am really stretching on this. Maybe it has nothing to do with the pain sensitivity tie-in. Maybe laughter just makes sex fun. As another anonymous Internet poster has said, "If you don't laugh during sex at least once, you're having sex with the wrong person."

WHY IS FOREPLAY IMPORTANT?

A SK YOUR PARENTS.

LAUGHTER HELPS SUSTAIN RELATIONSHIPS: MY CHAT WITH YAKOV

I'M NOT GOING TO GO INTO TOO MUCH DETAIL ON HIS BIO, WHICH is freely available as a matter of public record, but I will say that very few comedians have experienced the level of success that Yakov Smirnoff has enjoyed. Originally from the Soviet Union, he came to the United States in 1977 and began performing as a stand-up comedian. His act was unusual and he was a huge hit in the Cold War–era 1980s, when he appeared in several movies and TV shows including his own sitcom on CBS. Early in his career he was well known for a style of joke we call the "Russian reversal," but he has long moved on from that format. As familiar with his work as I may be, I will always remember him best from his many appearances on the show *Night Court*, one of my all-time favorite shows.

I previously mentioned that he now owns a theater in Branson, Missouri, where he continues to perform and that he has since earned a master's degree in positive psychology. His show *Happily Ever Laughter* uses humor to convey his underlying positive message of how laughter helps sustain relationships. Being a comedian with a degree in psychology myself, I find this inspiring. I have often used a few of his quotes in my seminars, and I've included them in this book as well. When I heard that he was in Los Angeles as I was preparing to write this book, I reach out to him for an interview. I could barely contain my excitement when he agreed to meet with me over coffee.

We met at a Starbucks, between the offices of the Screen Actors Guild and the La Brea Tar Pits. Sitting in the back, I tried my best not to act like a total comedy fanboy when he walked in.

Originally I had intended to use just a few bits and pieces of the interview to help add some new perspective to the subjects in this book. However, I had so much fun talking with him that halfway through the talk I realized that it would be better to present it as it happened. So here is my chat with Yakov Smirnoff, most of it anyway:

BK: Obviously, you have had an amazing career in stand-up comedy, and then you got interested in positive psychology. How did that come about?

Yakov: My interest was more specifically about laughter. I see laughter as a canary in the gold mine. In addition to being an expression of a very positive emotion of happiness, I also believe that it has the potential to let people know happiness is not there. It's a very subtle thing that sneaks up on people so they don't realize it until they're in some severe depression or in a divorce or in some kind of a condition that makes them obviously unhappy. But the process takes a long time. It doesn't happen overnight. That became my focus as I went through a divorce and I was realizing that the laughter was the thing that went first.

It's like there is a very specific order that happens to people. When they fall in love, there is a lot of laughter because it's a sign of happiness. It's an expression of happiness. That honeymoon stage continues and then they get intimate and they move in together and then they get married. Then, when things start falling apart, it's the reverse progression. Laughter goes first, second goes intimacy, third your house.

I also experience this: I have a theater in Branson, Missouri. I perform in front of four million people. I ask them, "how many of you experience laughter or remember laughter being a major part or behind-the-scenes of your relationship?" They all do. "Do any of you not have laughter?" Nobody applauds. "How many of you would like to bring laughter back to your relationship?" They all applaud.

BK: So you help them to bring back the laughter?
Yakov: First of all, I talk about this in my show and I find that people are fascinated because they recognize themselves within that process. They were doing certain things in their honeymoon stage that they are no longer doing.

Awareness is the number one stage of any learning. Once they're aware, then they're open. Then I share the differences between men and women, backed up with scientific knowledge. It makes people open up, laugh, and learn. That's why my whole show is called *Happily Ever Laughter*. That's the name of the show.

BK: You actually cite scientific research in your show?
Yakov: Yeah, I'll mention it. When the woman is sleeping, only 10 percent of electrical activity in her brain is shut down; the woman is sleeping 90 percent alert. They can Google it, and they'll see it's true. It's designed by nature so that in the middle of the night, she can hear what? They'll all say, "The baby crying." I'm saying only female voices just answered that question. Men, we don't know because 70 percent of electrical activity in our brain just shuts down when we are sleeping. At that point, we switch to natural gas.

BK: I like that.
Yakov: That is how I gift wrap that in laughter.

BK: Does your new show *Happily Ever Laughter* draw your regular comedy crowd or is this a new audience of people who seem to need the message?

Yakov: That's a good question. At this point, the new brand has not been seen on national television or anywhere. They're coming for the old brand at this point and then they discover the new brand during the show. But my intention in the next two years is to be able to introduce a new brand, the newer science of laughter. Then, hopefully, if that becomes popular with the audience, then they will be coming for the new brand, which will be open.

BK: I think it's fascinating that you found positive psychology, I really do. For me, psychology was my backup plan. I wanted to become a comedian before I became a psychologist, but I got this degree first. It's been nice that I've been able to bring the two together. From your perspective, is incorporating this world of positive psychology into your stand-up act a much more rewarding stage of your career?

Yakov: I think so. I think that I was doing it . . . in the eighties. I was helping Ronald Reagan put jokes in the speeches so they could help the Russian relationship. In some ways I was involved in this, but I wasn't aware of it. I wasn't doing it consciously. I was doing it just because I wanted to get nice car. Now, I feel like I'm doing it intentionally because I'm seeing how much pain is in the world.

If there is an awareness that can be brought to the world that can relieve some of that pain, it feels rewarding to me.

[Note: At this point Yakov was running short on time and so we had to wrap up our discussion. I asked him a couple more parting questions.]

BK: What would you advise people to do if they want to bring more joy into their relationships?

Yakov: Ultimately, I'm trying to give people an umbrella that will still be individually used by everyone differently.

I believe we created laughter in the honeymoon stage of the relationship, that we gave the relationship a lot of importance for whatever short period of time. There is research that shows about a year on average that we give [the relationship] importance and focus on each other's happiness. Then, we detour from that. Mission accomplished. I have a person that is in love with me. They said that they'll be there forever and I don't have to do this anymore. Just like saying, I planted a tree. I don't have to water it. It's going to be there forever. Mission accomplished.

I came up with basically focusing on that importance of the relationship and then fun and then time. I call it a GIFT: "Giving relationship Importance, Fun, and Time." That would be spelled out in the special, explaining the importance of giving versus getting, because that's the only control you have. You can give whatever you want. You can't necessarily get whatever you want.

Recognizing that, to give someone . . . If I want to create happiness, I need to give importance to the relationship, give it fun, and give it time. That's basically what we did. When I do it for me, when I just go, okay, I'm going to see my kids, how can I give that relationship importance, fun, and time? I think what's important for them? Certain foods, certain activities. . . . Then, what's fun for them? Totally different categories. They enjoy concerts. They enjoy music. They enjoy doing something at home. . . . Whatever. It's all in our memory bank. That's what we do when we are happy. Then, block off the time. Make that a priority.

BK: Excellent advice, thank you for sharing! The last thing that I am asking people is if you could give me an example of how you may have used humor to help you through some stressful event. Would you like to share a story with my readers?

Yakov: I'm sure I use it all the time. Just, if you're asking for specific story where I'm overcoming stress, some kind of stress, when I went for a divorce, there was a lot of stress, but now we're good friends. I don't send money to my other good friends, but they don't have attorneys in their speed dial either.

I'll give you another example. David Letterman had a top ten list of things that would change now that the Soviet Union is no longer there. I made number one on the list, "Yakov Smirnoff will be out of work." Think about stress there. I'm living in a two and a half million–dollar home in Pacific Palisades and, all of a sudden, all of my contracts in Vegas and Atlantic City, Reno, Tahoe were not renewed. Now, I start looking for a place where they did not know that the Soviet Union collapsed, I end up in Branson, Missouri. Twenty-five years later, I'm coming back to LA.

All of that is what you're talking about. I'm seeing that, yes. Does it release the stress to joke about it? Of course.

BK: Thank you so much for your time!

LAUGHTER HELPS CALM OUR MIND

THE FOLLOWING IS AN ACTUAL EMAIL I RECEIVED FROM AN AUDI-
ence member after one of my seminars. I have changed a
few words to help contextualize it, but otherwise it is verbatim:

> I found your lecture to be very helpful. I teach school and
> have a violent crazy neighbor who is focusing all of his
> angst on me. I find myself overreacting to most things and
> plagued by anxiety, so I am applying your wisdom to my
> life. Also, I was listening to the music you played in the
> morning and there was a song "Losers Love to Dance"
> that I can't find. Can you send me a copy of your song list
> so I can download some of them?

The song in question was actually "Lose Yourself to Dance" by
Daft Punk. If you are unfamiliar with the song, it was a pretty
big hit in 2013 and featured vocals by Pharrell Williams. The lyr-
ics are not that complicated. There are a few verses but basically
Pharrell repeats the phrase "Lose yourself to dance" over and
over for nearly six minutes. He is pretty articulate and it is a hard
line to misinterpret. Before my seminars I like to play uplifting,
positive music and I think that song definitely qualifies.

So one day in my audience, there was this teacher who
suffers from anxiety and claims to overreact to most things.

Someone who may or may not (depending on how much she is overreacting) be living next to a potentially violent person who seems to be angry with her. It is in this context that she hears a song with a fairly positive message and her brain perceives it as "losers love to dance." When I emailed her back with the correct song title she replied, "Wow, I heard that wrong. That speaks volumes about my psychopathology."

I share this exchange because I am very fond of stating that our emotional state is the lens through which we view the world. Coincidentally, shortly after writing this section, my friend comedian Kendra Cunningham posted this joke to her Facebook: "I was a fat little kid. I always thought the game was called Duck-Duck-Moose." Our affect influences our perceptions. If we are feeling crappy, then we are going to be overly focused on the negativity in our lives and our emotional lens is going to make us misinterpret the positives. If we are feeling anxious or fearful, that is going to have a similar impact on our perception of our experiences. You might have heard the expression that when someone is overly happy or optimistic they are looking at the world through "rose-colored glasses." Well, the opposite can be true as well; there are plenty of people who look at the world through a pair of glasses smeared with shit.

So, how did we end up with a pair of excrement-covered spectacles? Most of us start out with a decent pair and they get tinted or dirtied as we use them. In my introduction I suggested that for most people, emotional disorders, including anxiety disorders and depression, are partly behavioral in origin. Much like how the choices we make throughout our lives impact our physical health, our cognitive and emotional choices impact our emotional health. Our thoughts and emotional responses are just decisions that our brain makes in response to stimuli. Unfortunately, as a result of a lifetime of poor choices regarding our emotional health, some of us have developed an emotional disorder and accidentally covered our glasses in feces.

Perhaps these choices are made on an unconscious level, and most of them are; however, we need to recognize them as such if we have any desire for change. Take worry, for example. Nobody just decides to worry. Nobody has ever said to themselves, "Wow I am so bored. I really feel like worrying right now. What should I worry about? I know, my kids haven't called in a while, let me worry about them." I am pretty sure that has never happened and yet plenty of people worry. It is an automatic response that is decided upon unconsciously but then consumes our conscious thoughts. Worry is a big contributor to anxiety and depression—so are pessimistic thoughts, guilty thoughts, and being judgmental or just negative overall. These thoughts and the decision to have them are not made by your prefrontal cortex; the choice is made unconsciously. The first step to improving is to recognize this and be more consciously aware of the bad decisions your brain is making.

Our thoughts can make us crazy. It can be the other way around (our crazy influences our thoughts), but for most people suffering from an emotional disorder, the negative thinking was a contributor before it became a symptom.

Recognizing the impact that our behaviors, thoughts, and emotions have on each other is the basis of one of the most effective forms of psychotherapy ever developed, cognitive behavioral therapy (CBT). Without going into too much detail, in CBT we focus on changing our unwanted and unhelpful thoughts and behaviors to help modify our emotional state. Based on the discussion so far, you can probably guess that CBT is very useful in overcoming emotional disorders. CBT was originally designed for the treatment of depression, but has also proved very effective in helping people overcome anxiety disorders. Eliminating or reducing unwanted thoughts results in a general calming of the brain. When we are calm, we are less reactive to stress, and as I have mentioned, most of our stress comes from our own thoughts.

In my opinion, the poster children for "mental calmness" are the practitioners of Buddhism. Buddhists seem very at peace with the world, and indeed that peace is part of their teachings. I am not alone in my opinion, and many other scientists have been interested in studying Buddhists for this reason. Specifically, scientists have been interested in a traditional exercise regularly practiced by Buddhists called meditation—more specifically, "mindfulness." Perhaps you've heard of it? Mindful meditation of course is not exclusively practiced by Buddhists, but they did develop many of the techniques we use today more than six thousand years ago.

The fact that there are scientists out there researching the impact of meditation is indicative of a pretty big paradigm shift over the past few decades. Previously, meditation was considered a subject outside of the realm of scientific interest. It was lumped along with a bunch of other seemingly belief-driven practices that just didn't have a place in science. I mentioned positive psychology was born in the 1990s, when psychologists started putting more emphasis on positive human functioning and happiness. One of the more interesting aspects about meditation is that for thousands of years, meditators have been claiming that the practice makes them happy. Therefore given our sudden interest in happiness, it was only a matter of time before researchers started testing these claims.

And it is a good thing they did because yes, meditation does indeed make us happier.

Being mindful is often described as being able to be fully aware of our thoughts, feelings, and sensations from moment to moment, without necessarily acting on them. It is a state of awareness that is achieved through practice and as I mentioned, it makes us happier. This is not just a subjective assessment of happiness either; mindfulness meditation actually increases the activity in the part of the brain that we associate with happiness (part of the frontal cortex). It also helps us reduce stress both by decreasing the activity of the amygdala so that we are less reactive to things that are not really threatening, and by increasing our sense of control thereby allowing us a greater opportunity to remain calm when we are threatened. It also helps people become more resilient, so that they recover from tragedy faster, and over time can increase their conscious awareness of physiological processes.

As effective as CBT is for treating depression and anxiety, it may even be more helpful to add mindfulness meditation to the mix. Learning how to meditate, and of course regularly practicing meditation as a part of therapy can further help an individual overcome negative thinking. This is the basis for Marsha Linehan's dialectical behavior therapy (DBT),[85] which, in very simple terms, is cognitive behavior therapy plus meditation. As you might expect, it seems to be quite effective at treating emotional disorders as well.

In the interest of full disclosure, I was one of those researchers who felt that meditation was not an appropriate subject for science. As I've shared, I began my career in neuroscience studying rat brains. My peers and I considered what we did as "hard science" because our laboratories were filled with cool equipment and reeked of chemicals. We wore lab coats and goggles and ran rats through mazes, occasionally dissecting a brain or two. You know, HARD science. We were distinguishing ourselves from others in our department who practiced more soft science. Their labs had comfortable couches and refrigerators that were stocked with bottled water. They asked people about their "feelings" and almost never ran one of their participants through a maze. They often had children come into their laboratory for testing, and had a collection of toys. We were hard and they were soft. And we were so mistaken.

This distinction was mostly an inside joke in my department, but it also reflected a fundamental difference in our approach to the science. I can't speak for everyone but I believed that the key to understanding or modifying our thoughts or behaviors would come from increasing our knowledge of biological mechanisms.

By learning more about the complexities of the brain and how it interacts with other systems, we would eventually know all we needed. The hard science focus made me biased against certain topics, like meditation. To my focused mind, the prospect of studying the effects of meditation seemed just as unappealing as the prospect of studying the paranormal or magic.

Part of that bias came from what we knew about the brain already. I had been taught, and later taught others, that once the brain developed to adulthood it was incapable of changing. I wasn't purposely spreading depressing misinformation; this is how we understood it. When a person is born, their brain is not fully developed. It continues to grow throughout our lives until early adulthood, when it is finished. Once a brain is done, that was it (so we thought). The only change we could expect to observe would be the deterioration some of us experience later in life. If you knew that the brain was incapable of change, then what would the point of therapy be? Why not just work to develop a better pill? You can see how thinking the brain was relatively static at adulthood would stifle interest in any practice that might claim to change it.

Just to be clear, we now know that the adult brain is definitely not incapable of change. It is very resistant to change sure, and most of us do not experience substantial change throughout our lives, but it absolutely can be changed. Neuroplasticity is your brain's ability to modify itself as needed based on the changing demands of life, and our brain has a tremendous capacity to do just this. We did not always have the understanding of neuroplasticity that we do today. We now know that we can, through hard work and conscious effort, impact the very structure of our brain. Geeks call this neuroplasticity, but you might call it "learning." So I have seen the light and I now recommend meditation on a regular basis, to anyone, not just people who suffer from disorders. I tour all over the country and I always recommend meditation, because the potential benefits are pretty amazing and the risk is ridiculously low. I already listed a few benefits, but did not yet mention that to see an impact research shows that we only need to meditate for about ten minutes at a time, twice a day. Who can't find twenty minutes out of their day?

There is an old Zen saying: "You should sit in meditation for twenty minutes a day, unless you're too busy. Then you should sit for an hour."

Because I don't like to recommend things that I am unwilling to try myself, two years ago I signed up for my very first meditation retreat, something that the hard scientist in me would never have done. For about two weeks, I voluntarily went into the desert of Southern California, took a vow of silence, and practiced meditating. With no contact with the outside world, I was left with only my thoughts to entertain me and I can say that I went completely out of my mind. It seemed that every negative thought I had ever had and stuffed into my unconscious came to the surface. I revisited every bad memory, every failed relationship, and every regret, and they haunted me as I struggled to stay focused on clearing my mind. It was a terrible experience and there was not a single day that I did not think about leaving. Every day I was there was the worst day of my life and worse than the day before it. After a while, I could swear that I actually felt my frontal cortex hurt. It was an absolute hell, and I still can't believe I endured it. I hated every minute of it. But I would totally go back.

By the way, I do not recommend that a first-time meditator try such an intensive introduction. I did so because I'm the kind of guy who jumps into the deep end of the pool, naked. Also, I figured that I would probably not continue with the practice for long, so if I was to experience anything it had to be an intensive introduction. I figured right; I haven't meditated much since that retreat.

At this point you might be wondering, "What the hell does any of this have to do with laughter?" and I don't blame you. I do seem to have gotten a little off topic, but let me tie it all back together.

The reason that I chose not to meditate much after my retreat is that after having experienced it firsthand, I feel that the benefits are very similar to those received through laughter. I believe that having a sense of humor helps scrub the brain of unwanted thoughts, reducing the likelihood of both anxious and depressive emotions.

The importance of laughter is not lost on meditators. There is another Buddhist proverb that I would like to share: "If you are too busy to laugh, you are too busy."

I have already discussed how laughter makes us happy, and it definitely does that. However, depression is a disorder that is not always associated with negative affect. The impact of regular laughter goes beyond giving us something to cheer us up once in a while, but it helps keep our negative thoughts in check. I also went into detail about how laughter helps reduce stress. Anxiety is basically our stress response but instead of a bear attacking us, the stimulus is inside our head. Again, the impact of regular laughter goes beyond momentary stress relief; it can help prevent our brain from generating the anxiety-provoking thoughts in the first place.

Obsessive-compulsive disorder (OCD) is an anxiety disorder characterized by ruminating on intrusive thoughts or images that typically drive an individual to engage in some ritualized behavior pattern for relief. Repetitive hand washing is the example most people are familiar with, but there are people who suffer from OCD who feel they have to do things like walk a certain way, perform tasks a certain number of times, or obsessively clean or count objects in their environment. Occasionally, it can even manifest with no visible compulsive behaviors. For example here is how comedian and writer Abby Rotstein describes her OCD:

I have what is known as "Pure O" OCD, or purely obsessional OCD. Essentially, doctors used to believe that those with Pure O didn't do the compulsions. We do; they're just all in our head. We can be having a conversation about tuna fish sandwiches while in my head I see visions of my family members dying or contracting some visually awful disease. I like to say I have a nasty thought bubble that follows me around, but no one's the wiser.

I wanted to introduce Abby because as someone with an anxiety disorder, she is familiar with mindfulness meditation and its usage. And as a comedian I find her perspective interesting. She went on:

> I have OCD and use performing as a way of coping. There's a lot of talk about mindfulness meditation and how it's beneficial for anxiety disorders. I think stand-up is its own sort of meditation. This might sound odd, but there are parallels. Mindfulness is really about staying in the moment. When I'm on stage, I have to be solely focused on the jokes or I'll screw up. In this way, performing is an escape, and for a few moments I'm someone without OCD.

As I mentioned, regular laughter and, as Abby suggested, performing comedy may provide our brain with a calming effect similar to that of mindfulness meditation. Now a word of caution, if you are a meditator I am not suggesting that you suddenly abandon what you are doing and try laughing more. Nor do I want to discourage you from starting to meditate if it is something you are considering. Meditation is a wonderfully beneficial activity and it only requires twenty minutes of your day. Remember, despite the fact that I don't practice it myself, I still tour the country recommending meditation as a healthy activity.

By the way, for those of you who may be struggling with OCD, Abby has a few pieces of advice:

Educate yourself about treatment options. Medication works for some, but ERP [a form of CBT] is more helpful for others. Find support, whether that's through seeing a therapist, attending group therapy, talking to loved ones, or through social media. Lots of people blog and tweet about OCD and anxiety, and it's so helpful to know you're not alone. Most importantly, tell your story. It's so powerful. You'll find people you can talk to, people who will share their own vulnerabilities with you. You might also help decrease the stigma surrounding OCD, and that may one day help others tell their stories as well.

And if you ever decide to try stand-up comedy, she says, "I think bombing on stage is nothing compared to living with terrible anxiety every day."

Mental illness, including emotional disorders, seems to disproportionately plague comedians compared to the general population. As a comedian myself, I can look around at the people I have met and worked with and this does seem to be the case, although I am unaware of any data that confirms this. Comedians do show greater incidences of some abnormal personality traits[86] and there are plenty of examples of comedians who unfortunately have taken their own lives.

Here is a funny side note: In preparing for this book I reached out to a lot of comedians battling various issues for commentary and interviews. Guess who got back to me the fastest? Those with anxiety.

During the summer of 2014 the world got substantially less enjoyable when we lost Robin Williams to depression. Robin was an amazing talent and a very sweet man, who impacted so many of our lives. When he died there was a lot of discussion about mental illness, particularly depression, in comedy. On tour, people would ask me, if laughter is so great for us and has all these benefits, then how would I explain what happened to Robin?

I would never get specific about Robin, or anyone for that matter, as it is difficult to know all the circumstances. However, if this is something that has crossed your mind in reading this book, I would like you to consider why we develop a sense of humor in the first place. I mentioned a couple of possibilities, such as relieving tension or using it as a defense mechanism. It seems reasonable to imagine that those with more darkness to defend against might be more likely to develop this skill. Even then, sometimes a good sense of humor is not enough.

I don't want to end this chapter like that, so instead I will share a story from Alan Moore's graphic novel *Watchmen*[87] that went viral on the Internet shortly after we lost Robin:

> Man goes to doctor. Says he's depressed. Says life seems harsh and cruel. Says he feels all alone in a threatening world where what lies ahead is vague and uncertain. Doctor says, "Treatment is simple. Great clown, Pagliacci, is in town tonight. Go and see him. That should pick you up." Man bursts into tears. Says, "But doctor, I am Pagliacci.

Whose job is it to make the comedian laugh? Hug your comedian today. Cherish the people who make you laugh.

WHY IS IT SO IMPORTANT TO OVERCOME NEGATIVE THINKING?

IN MY INTRODUCTION, I STARTED OFF BY MENTIONING EMOTIONAL disorders, and so it seems fitting that I would come full circle and discuss them in more detail toward the end. These disorders are relatively common; in fact they are way too common. Nearly a fifth of the population will suffer from an anxiety disorder at some point during our lives,[88] and 15 percent of us will suffer from depression.[89] These numbers have been rising over the last fifty years or so and I find this very troubling.[90]

Emotional disorders are awful. I have been personally affected by both depression and anxiety in many ways throughout my life. I have directly experienced a few bouts of depression; for example as I mentioned previously I was in a miserable state when I first took the stage as a comedian. I have had family members and close friends suffer from the disorder. Professionally, as a psychologist but even as a public speaker, I have encountered hundreds of people who deal with depression and anxiety. You can imagine that I also encounter these disorders quite a bit as a comedian. They have a significant impact on not only the person directly suffering from the disorder, but also on the people closest to that individual and society at large.

Our methods of treating depression are medication and therapy. That's basically it. On the road people always ask me for alternatives, and I'm not sure why. I remember one woman said something like "I'm in therapy, and my doctor wants me to start taking medications, which I don't want to so what else would you recommend?" The only thing I recommend is taking your damn doctor's advice. If your doctor recommended a cast for a broken arm, you wouldn't ask, "yeah, but what else?" I've never understood questions like that. There seems to be a strong anti-medication belief held by some people (which is not at all ironic since they are being treated for a disorder significantly influenced by their beliefs). I understand that certain medications have undesirable side effects, but the direct effects of the illness are potentially WAY WORSE. When a part of your body is not functioning properly and medical science has a solution, you take your damn pill. Imagine if people suffering erectile dysfunction had the same attitude; "Gee, Doc, I get that you say I should take this little blue pill to achieve an erection, but what else do you have?" You'd never hear that, that pill would be swallowed and Grandma would be smiling before the doc finished writing the prescription.

The resistance to medication is especially confusing when you consider that people are even more resistant to a behavior change. As I've discussed, depression and anxiety disorders are a function of our behavioral choices and thoughts. To truly overcome the disorder, we need to change our destructive thinking. Therapy, such as cognitive behavioral therapy, can help us accomplish this. Medication may help relieve the symptoms of the disorder, but without real cognitive change we could be dependent on that medication for a very long time.

Let me be as clear as I can possibly be: People with these disorders need interventions that goes way beyond laughing more often or taking life less seriously. They need therapy and, depending on the intensity of their condition, potentially medication. I discussed the benefits of mindfulness meditation and made a parallel to the benefits of regular laughter, but these should never be viewed as substitutes for other forms of treatment. Meditation can help just like regular laughter can help. However, neither practice is a cure for anything.

Most of us do not suffer from an emotional disorder, but we can all benefit from a little mental calming. Learning to recognize the humor in life, taking it less seriously, and regular laughter helps train our brain not to worry so much or ruminate on negativity, not to focus too much on guilt or regret, and not to be too judgmental. Just as I tour the country discussing the awesome benefits of meditation and recommend the practice to anyone and everyone, I also recommend developing that sense of humor.

LAUGHTER GENERATES
MORE POWER THAN SCREAMS

According to researchers at Monsters University, making a child laugh generates at least ten times more gigglewatts of power than making that child scream in terror. This research,[91] led by noted Monster scientist Mike Wazowski, has resulted in tremendous social change in the city of Monstropolis, as well as much relief to human children.

THE BENEFITS OF LAUGHTER: A RECAP

HEY GANG, WE MADE IT! I JUST SPENT NEARLY HALF OF THIS book discussing the various benefits of laughter, and you were right there along for the ride. I feel like we should celebrate, who is buying?

I imagine that I write slower than you read, so it probably feels like more of an accomplishment to me than you. But whatever, I don't see your name on the cover so I am going to call it a milestone and get myself a cold one.

Okay now, let's review all the potential benefits that have been associated with laughter.

First, I mentioned that laughter makes us happy. This is my favorite benefit of laughter, because if there is one thing I appreciate, it is being happy. Happiness is great, and it feels good to be happy. Speaking of which, laughter feels good. Laughter stimulates activity in areas of our brain that are associated with pleasure and increases the production of our endorphins. Endorphins provide us with pleasure while at the same time decreasing pain. Laughter also decreases pain. This is mainly thanks to those endorphins but laughter also distracts us from pain and reduces our anxiety or stress.

Oh yeah, that reminds me that laughter reduces stress. There are a lot of benefits associated with laughter, but I believe that stress reduction is the most important. It is difficult to be happy when we are stressed and stress contributes to a number of physical and mental issues. For example, stress decreases the effectiveness of our immune system. Conversely, laughter has been shown to boost our immune system. Stress can contribute to heart disease, diabetes, and cancer. However, laughter is associated with lower blood pressure, lower blood sugar, and an increase in the activity of cancer-fighting cells in our immune system. Stress is also a major contributor to the development of depression and anxiety, and laughter helps provide relief to both. Short-term stress may be useful, but prolonged stress can be very detrimental to our health. Reducing stress is healthy. Exercise also helps reduce stress. Laughter is also a form of exercise. It is not usually strenuous exercise, but it can be.

Laughter is also inherently social, and is very important to establishing good social bonds. The hormone oxytocin is released when we laugh. Oxytocin helps suppress our stress response. It is also released when we are nursing, hugging, or making other positive physical contact with others (you know what I mean ... brown chicken brown cow), helping us to establish stronger emotional bonds. Laughter helps us turn strangers into friends and friends into lovers. Other people make us happy. Sometimes, we have sex with them. Oxytocin is also released after sex, and laughter is sexy. Sex also helps relieve stress. In general, men and women are attracted to partners who make them laugh. Smiling and other displays of happiness are also considered more attractive. Laughter makes us happy. . . . And that brings me back to doe, a deer, a female deer ...

I hope that was not too confusing for you to follow. I chose that approach instead of listing out all the benefits into bullet points because I want to make one additional point before I move on: All of these benefits are interrelated. In research, we need to control for certain variables or narrow our focus but you should understand that these benefits are not independent. This makes the argument for laughter much more agreeable.

OKAY, SO WHAT CAN WE DO
WITH THIS INFORMATION?

WHEN I FIRST OUTLINED THIS BOOK, I ENVISIONED IT LIKE ANY good piece of entertainment, playing out in three acts, each one answering an associated question: "What," "Why," and "How." In the first act, the "What," I introduced you to the main characters, Humor and Laughter, and their little buddy, Jokes. I discussed how humor is basically anything we find funny, and that our brains interpret humor based on recognizing benign violations. I described jokes as an action with humorous intent. Regardless of whether our audience is amused, jokes are an attempt to elicit laughter from someone. A joke can be any sort of behavior, but most of us are familiar with verbal or written language–based jokes. Then we got to know laughter and understood that as our body's reaction to humor, it can range from a very subtle response to a rolling-on-the-floor, pants-pissing, uncontrollable fit of activity. In the second act, the "Why," our characters began their story. Humor and laughter, hand in hand, gaily skipped all over the world making people happy and helping them feel good. In their path, they had an encounter with their enemies, stress and anxiety, and were triumphant in calming the brain and helping the sick people feel better. Knowing that these enemies would be back again for the sequel, they then started exercising. They fell in love, and did the nasty. A lot. Seriously, get a room, you guys.

Now here we are at the third act, the "How." We've had our story arc, and now we are looking for a resolution. How can we use this information to benefit our lives or those of our loved ones? That's not a bad question, and I am eager to find out how this thing ends myself. Not that I'm in a hurry to get out of here, I love you people.

I have mentioned this a whole bunch of times throughout this text but it is worth repeating again: Laughter is not a cure for anything, except maybe a temporary case of the blahs. Laughter is not going to eliminate our pain, improve our health, remove our mental hang-ups, cure our cancer, or lance that wart. The best that we can hope for is that laughter makes life better. If there is something, making our life shitty, laughter can improve our situation, not fix it.

You may recall that when I was discussing stress, I mentioned being in traffic. Some people get really stressed out in traffic, sometimes to the point of developing road rage. I do not find traffic all that stressful. I am not saying that I like traffic—in fact, I pretty much hate it—but it does not stress me that often. To me, traffic is just cars moving slowly. Besides, heavy traffic is the perfect time to send that text message you've been putting off or update your Facebook status. Okay, just in case this entire book has been lost on you or you are from the Midwest, that last line is absolutely a joke.

I bring up this example again because most of the wisdom we can gain from the research on laughter points to its ability to impact our lives by impacting our attitudes. Being able to make a joke about a boring, painful, or desperate situation has a way of cushioning the blow. Our attitude toward life is extremely important to our health. Think about how much less aggravation we would suffer if we had a more positive attitude toward something like traffic. We would drastically cut down on anger and distress; we would reduce our likelihood of engaging in risky behavior like tailgating or flipping our middle finger at the wrong person. If you don't think raising your finger is risky,

I once had a window on my car smashed in while stopped at a red light in Texas, because my backseat passenger felt the need to display his finger (I am guessing he was proud of it) to the angry man in the car behind us. And I think my father only got out of the habit of flashing his when he was stabbed in the gut on his way home from work one night. A more positive attitude toward traffic would have prevented both of those interactions.

Also, if you are in Los Angeles, it also helps if you avoid all freeways and don't even dream of taking the 101 during rush hour or the 405 ever. While I'm at it, and because I live in West Hollywood, just go ahead and quit taking Sunset Boulevard too.

Now, if you are one of those people who is easily aggravated by being in traffic you might be asking, "How do I develop a more positive attitude to frustratingly sitting in my car, boxed in by other cars, and surrounded by jerks?" First, I would like to remind you that this is a book and I can't hear you, so talking to the pages is only going to make you look even more insane. I also can't tell you exactly how to calm your overreacting brain in this or any context (generalize!). However, I can give you a few tips to try.

First, remind yourself that things are going to be okay; they will work out one way or the other. Then, try redirecting your thoughts to a different subject. In other words, think about something else—preferably something positive. Distract yourself with music, if you aren't already listening to some. Distract yourself with a phone call to a friend, provided, of course, you are using a hands-free device because, duh, we all use those, right? Breathing helps reduce our stress response and decrease anxiety too, so take some deep ones. This is generic advice, and can apply to any stressful situation.

With regard to using humor in a stressful situation, such as being stuck in traffic, I find it helpful to laugh at my circumstances. Sometimes, you just have to laugh, and being worked up about a traffic delay is damn ridiculous when you look at it objectively. Laugh at the fools behind you who think honking their horns is going to have any effect, as if all the drivers in front of you simultaneously forgot how to accelerate and need prompting. Laugh at the absurdity of the situation. Make a joke to yourself, not just in your head but actually say it aloud. Even if it doesn't make you laugh, it will help to alleviate some of the stress. If you aren't good at making jokes, I give a few simple techniques in the next section. And finally, think about something funny: a past experience, a movie you enjoyed, a joke you might have heard. Think about anything to get your mind off of the fact that you are running late for a meeting with your parole officer.

Among other things, humor helps promote optimism. Optimism in turn helps make us happier and more resilient and if you think about it for a second, that makes perfect sense. Optimistic people believe their situation will improve. When I am in traffic, I know it is only temporary and I will get to wherever it is I am going; then again, I am a naturally optimistic person. I also know that traffic is outside of my realm of control, so there is nothing I can do if I am late for something. Recall that I also discussed the value of feeling in control with regard to our tolerance for stress. Sometimes, accepting that we do not have any control is the more comforting option. (If there is nothing you can do about it, why are you worried so much about it?) When I am sick or in pain, I know I will get better. When I was suffering the horrible stress of graduate school, I knew that eventually I would get a degree. When I was broke and unemployed, I knew that those situations would improve as well. Optimism helps sustain us and humor can make us more optimistic.

Some say the glass is half empty; some say it is half full. I say, why are we calling a red plastic Solo cup a "glass"?

Forget the glass, where's the keg?

United States senator and presidential runner-up John McCain attributes his ability to laugh and his natural optimism to helping him survive his time as a prisoner of war in Vietnam.[92] He has stated that his positive attitude was one of the things that helped him come out of that horrible experience with a relatively good mental state. He was tortured, put in solitary confinement for several years, and yet held onto the belief that things would get better (spoiler alert, they did). He also reported there were many times when he felt happiness in that prison, and would LAUGH at his guards. Now that is a maverick.

Besides, he selected Sarah Palin as his presidential running mate in 2008, and that just HAD to be a joke.

Having a positive attitude is difficult for a lot of people, especially those in pain or suffering from an emotional disorder. I have presented the possibility that our attitudinal choices contribute to our pain and help us develop those disorders, so it is important to shift our thinking and our behaviors as soon as we can. We can reduce our risk of developing a disorder, and we can overcome the various disorders we find ourselves facing, but it is not going to be easy. For many of us, that means we are going to be attempting to override unconscious decisions to react negatively to trigger stimuli. We are going to have to use our wonderfully complex prefrontal cortex to suppress our immediate gut reactions to threats misinterpreted by our amygdala. In other words, we are going to have to make some serious, conscious effort to not take life so seriously. Seriously, we need to . . .

LEARN HOW TO NOT
TAKE LIFE TOO SERIOUSLY

THIS IS MY CENTRAL THEME BUT I WILL BE HONEST HERE, I REALLY
have no idea how you are going to accomplish this. I
believe that I have made a case for its importance, presented
a great deal of benefits associated with laughter, but if you are
one of those people who takes life too seriously I cannot offer
you any easy advice. Actually, here is some easy advice: GET
THERAPY. But be prepared because change is going to take a
lot of time, be difficult, and require a lot of effort, and you may
not be entirely successful. However, there are a few things that
I do recommend we internalize from reading this book.

First and most importantly, we should learn to recognize humor. Humor is all around us, and almost anything can
potentially be a source of humor. I described benign violation
theory as a means to explain almost every example of humor
we may encounter. To review, according to BVT, humor is the
result of our brain initially perceiving some stimulus as a threat
or contradiction to our expectations, and then recognizing its
nonthreatening or benign nature. Our brain realizes its mistake
in its initial assessment and we respond by finding the situation humorous. Let's use one of the most basic examples, tickling. With tickling, someone is touching us, perhaps in a part
of our body that doesn't normally receive a lot of attention,
like an armpit, and so our brain initially perceives that touch as

an assault. The person doing the touching is most likely someone we know and trust, so our brain reassesses the situation as
benign and it makes us laugh. In the same way, our best friend
might say something like, "Hey, douchebag!" and we might
laugh it off because hearing an offensive greeting come from
a good friend is clearly a benign violation. We don't have to
laugh, but we probably don't need to punch that punk ass bitch
in the mouth either.

Understanding what makes things funny is the first step
to understanding jokes. Jokes are just actions with humorous
intent. They are not always funny, but they at least are delivered with the intention of being funny. Having that piece of
information in our back pocket can help us reduce interpersonal conflicts and personal insecurities tremendously. It can
help us get along better socially, and maybe even create a few
jokes of our own.

I would like to point out that in the prior paragraph I am
thinking of the importance of understanding what makes things
funny from the point of view of an observer, the joke recipient.
If we are the ones making the jokes, it is even more important to understand humor. I have observed many people, from
aspiring professional comedians to the loudmouthed guy at the
bar, make jokes that not only failed to make anyone laugh but
also resulted in offending the intended audience. Generally,
jokes make us laugh and increase joy, but a bad joke can have
the opposite effects. People can use humor maliciously, to insult
or hurt people, and those people are just assholes. My point
here is that if you are not an asshole, but are consistently telling
jokes that people are either not laughing at or at worst, getting
offended by, it is equally important to understand what makes
things funny. If you find yourself saying, "Eff them if they can't
take a joke" more often than you'd like, the problem is not the
audience.

Every joke will eventually offend somebody. That is a basic
rule of comedy. You might remember my joke about eating

cookie dough for breakfast. The last line of it usually goes "Now, do you skinny people know another way to eat cookie dough that I'm not aware of? You take a bucket of dough, a spoon, and some depression . . . that's a meal!" Again, it probably does not read well, but I assure you when delivered live, it almost always gets a good laugh. Regardless of whether the joke bombs, I feel like it is pretty innocuous. From my perspective, there is nothing in that joke that should offend people. The joke is about me, and my ridiculous, and mostly fictional, dietary choices. That is it. Yet one night after a show someone came up to me and told me that the joke had offended her. Apparently, she thought that through my joke I was stating that all people with depression are going to have problems with food. Which is absurd, and no way implied by my joke, but like I stated every joke will eventually offend somebody.

The jokes I've included in this book are bound to offend some of you, but if you've gotten this far, then you are probably doing okay.

Very few people have a desire to be lonely—alone maybe, but not lonely. This reminds me of a story from my childhood. Before I found other activities that I enjoyed, I attempted a few sports. Well, "attempted" probably isn't the right verb; against my desires, my parents signed me up for a few sports. I was sucky at all of them, including some of the easiest games imaginable: T-ball and bowling. If you don't remember T-ball, it is the same thing as baseball but instead of having the ball served to you by a pitcher, it sits on top of a tee and the batter just swats at it. I had such bad hand-eye coordination that with almost every swing, I would fall short of the ball and bang that bat into the side of the metal tee, sending painful vibrations up the length of my arms. Doing so was the equivalent of a strike, and I would strike out all the time. At freaking T-BALL! That's like falling over on your bike with training wheels. Anyway, I had such a miserable time with it that eventually I gave up trying and just went up to the plate anticipating that I would strike out. When

I did, I would act as if I intended to do that and resume my spot on the bench. It was a defense against the negative emotions of failure. I feel like assholes develop their style of humor in a similar way. They may have started out trying to tell a joke to fit in socially, but failed. Then, after failing so often, they adopt an "I meant to do that" attitude as a defense mechanism. If we are accidentally pushing people away because every time we say something we offend them, then yes, we should be evaluating our understanding and use of humor.

Understanding what makes things funny is also the first step to identifying the humor in our lives. There is humor all around us, and people manage to laugh even when suffering some pretty dire circumstances. Throughout this book I have given you plenty of examples of people finding humor in tragedy. These examples have ranged from dealing with the stress of being a police officer, to dealing with chronic pain and illnesses like diabetes and cancer, to coping with depression and anxiety. For even more examples, refer to the "Humor Testimonials" chapter at the end.

I will give you one more example. My girlfriend recently treated a diabetic who unfortunately was having his leg amputated. He was in remarkably good spirits for someone about to lose a limb and when she asked why he was in such a good mood, he said, "Think of all the money I'm going to save on shoes!"

If you just got diagnosed with diabetes or cancer, I can't tell you how to find the humor in that but I can tell you that I bet it starts with finding the humor in traffic. Not traffic specifically, but hopefully you get the metaphor. Being able to find humor in the small stuff builds our ability to find humor, and as a result stay positive, when facing the big stuff. I have always been a fan of Richard Carlson's message "Don't Sweat the Small Stuff"[93] because ultimately, with the right attitude, it is all small stuff. Humor is many things, including a very important coping mechanism.

I want to share a story with you about a very strange encounter I had on my lecture tour. First, you should know that when I am on a speaking tour, I do a seminar every day Monday through Friday. Every day, I am in a different city and then I travel on to the next city. The schedule makes it hard to keep up with dry cleaning, and I try to travel with minimal luggage. My seminars are informal and a full day long, so when I present I like to be comfortable. My usual outfit of choice entails my throwing a sport jacket over a button-down shirt and finishing with a nice pair of jeans. It's a uniform that fits my personality too: not too casual, but also not stuffy. Anyway, one day in Houston during a break a woman came up to me and said that she was personally offended that I would wear jeans to my seminar. She was furious and full of anger at the very nerve of my unprofessionalism. She said, without any hint of irony, that my jeans were so offensive to her, so distracting, that she could not pay attention to my seminar on humor and happiness. How dare I! It took a lot of effort for me to refrain from saying aloud something like, "Lady, this is a seminar on humor. I am not here to sell you real estate."

Besides, it was Texas and I was wearing cowboy boots. That should have at least counted for something. And, it was hot as hell that day. She was lucky I had on pants.

Clearly she needed the message, although I doubt she allowed herself to hear it. She was taking life, and obviously her expectations of a comedian's wardrobe, way too seriously. My response ("not here to sell you real estate") was my brain using humor to minimize the potential threat she represented and therefore not get offended by her absurd accusations. Although I held back verbally, my brain did this automatically because it knows not to take things too seriously. Besides, it was my seminar and I can wear whatever the hell I want. If she needs a clean-cut guy in a three-piece suit to tell her how to be happy, then she needs more than a one-day seminar.

Whatever you are worrying about right now isn't worth it. I don't care what it is, it is not important enough to worry about.

Developing our sense of humor, and learning how to take life less seriously, can be an important step to improving the quality of our lives. In this book, I've discussed the elements of humor and how to recognize them, and a bit about joke telling, but some people may find it helpful to receive additional instruction. Which brings me to a point of debate among circles of comedians: can you teach someone how to tell a joke?

LEARN HOW TO TELL A DAMN JOKE

Comics argue this all the time, especially when the topic of instruction comes up. Can you teach someone how to be funny?

You might be able to pick up a book of street jokes and memorize them. However, I do not think a sense of humor can be acquired through training, but we can learn to develop our existing abilities. I mentioned that I began my comedy career at the San Francisco Comedy College. The instructors there were all professional comedians; some had toured the country for years whereas others were based locally. They did not teach me how to be funny, but instead taught the basics of joke writing and performance. They drew on each student's natural sense of humor, and helped us to recognize the important elements of a good joke, communicate more effectively, and establish that important relationship with the audience. When I entered the greater comedy community, I felt as if I had skipped years of open mics thanks to the instruction I received. There are a lot of comedy classes out there. Most comedy clubs offer their own classes, and in cities with a heavy concentration of entertainers (like Los Angeles, duh), they seem to be everywhere. I have heard good things about some and heard of others that were not so great; there isn't any accreditation body for comedy instruction after all, so I figure I got lucky in San Francisco.

I believe that you cannot teach someone to be funny, but you can teach basic joke structure. So for those of you who might need a quick lesson in joke writing, here are two easy ways to create a joke. Remember that humor is based on violating an expectation, so both of these formulas capitalize on that. One: take a word or phrase that has multiple meanings and set up the joke by leading your audience into one meaning, then revealing the other through the punch line. For example: *I went to my family picnic recently, it was nice but there were too many ants. Yes, my mom's sisters were in town.* That's a bad joke, but I took about five seconds to think of it using the rule I just shared with you. Keeping that formula, you can probably crank out dozens of such jokes and maybe one or two might be funny. The second involves what we call "the rule of threes." Anytime you make a list, whether you realize it or not, you are setting up an expectation in the mind of your audience. The first items set the pattern; the last item contradicts the expectation. Three items in the list seems to be the magic number for tricking the brain in this way. For example, with the exception of my current girlfriend, Sarah, I have a strange attraction to women named for places in the South. I have dated girls named Savannah, Charlotte, and Hilton Head Island.

See what I did there? Hilton Head Island is clearly a man's name.

Here's a third, bonus quickie for generating laughter. There are basically two ways to make people laugh: we can say something funny (and the two formulas above will help with that) OR say something in a funny way. Voices and accents are great for making something otherwise mundane words into a great joke. If you can say something funny in a funny voice, then you have a career. I have mentioned comedian Gabriel Iglesias a few times. Gabriel uses a lot of voices in his act. So do Jim Gaffigan and Jeff Dunham, and even if you have never heard of comedy before, I'm pretty sure you have heard those names. I have one voice, my mother's. It is an exaggerated high-pitched nag in a New York accent, and if you ever meet her you'll realize it is spot on.

Side note: after a recent show in Las Vegas, a guy came up to me and said, "I love all the voices you do" to which I replied, I only do one voice, but I do it a lot.

The life of a stand-up comedian is not for everyone, and I would not recommend it at all. It can be, and sometimes is, extremely rewarding. But it is also horribly frustrating and poorly compensated. In fact it usually takes years before a comedian even starts getting paid for gigs, and when they finally do start to earn a living, it's a meager one at best. Fame is a long shot, and the odds are stacked way against us. It is not uncommon for full-time, professional comedians to still spend nights in their cars or couch-surfing, or to temp during the day. We make it work because we love it.

However, not everyone who takes a class on stand-up comedy wants to be a comedian. Lots of people take classes to learn the basics of humor, and the types of things I've mentioned, but there are other benefits. Stand-up comedy training lends itself very well to the world of public speaking and communication in general. As I discussed, I was a professor in a prior life, and as a consultant I gave many presentations to various clients over the years. All of this experience helped prepare me for my career in public speaking. However, it was my training in stand-up that really made me made me a better speaker and a more effective communicator in general. One of the biggest fears people have is public speaking, and another is the fear of being laughed at. Learning to perform stand-up comedy helps eliminate both of those fears. You don't have to want to be a comedian to benefit from taking a comedy class.

But, let's just say hypothetically that you do want to give comedy a shot. It is important to mention that just as laughter has been shown to be quite valuable, there are some real benefits to performing comedy as well. If you've been paying attention to the comedian interviews I've included throughout this book, or even, um, the content, you have already picked this up. Humor does not only benefit the audience. Ask any comedian and they will probably tell you that performing has a therapeutic benefit. Ask me (cuz, you know, I am one of them) and I will tell you that there are few things that feel as good to me as stepping off stage after killing it during a set. That is such a high. We get to share our misery, express pain, and air grievances. The audience laughter and applause (when we get it) provides us with positive reinforcement and elevates our mood.

I mentioned earlier that if you follow comedy enough, you will notice the pattern of comedians bringing their pain to the stage. For a lot of comedians, the stage is therapy. Some comedians will literally state that as a hacky punch line, but it is true. As far as I am aware, the therapeutic benefits of performing have not been empirically researched, but there is a substantial amount of anecdotal evidence.

It may help reduce pain. It does for comedian Trevor Harris, who told me that he was "surprised and intrigued" when his doctor suggested that he try stand-up comedy for his pain. On how laughter and performing help to manage his pain, he said:

> I noticed all my pain starting to dissipate. As soon as I started laughing it was like pressure from a bear hug that was too tight was releasing. Eventually, to where I felt like I was floating. Same for when I first tried out stand-up. I was seriously nervous at first. But when I got my first laughs, I started to loosen up a little and [was] starting to become more comfortable on stage with the adrenaline rush. It didn't matter whether I was good or completely sucked, the pain relief kept happening during and after sets for up to a week and a half after each one. . . . Stand-up comedy has been I think my main tool in the healing process along with some pain killers on the side that I'm now in the process of getting off because I'm doing so well!

For more about Trevor's story, check out his testimonial at the end of this book.

I am convinced that there is therapeutic value in performing comedy, which is why I am so interested in the work of Canadian comedian David Granirer. David is not only a comedian, but also a mental health counselor who battles with depression himself. In 2004 he started an organization called "Stand Up For Mental Health[94]" and through it, teaches stand-up comedy to people with mental illness to help them build confidence. To date he has run the program in more than thirty cities, in Canada, the United States, and Australia.

I have always been interested in his work, so I spoke with David about his students. He said, "Some people are really high-functioning. They've got jobs and families and all that kind of stuff. Some folks are on disability. We've got every single diagnosis that you could possibly imagine." A few students take his course because they are interested in becoming stand-up comedians, but most take it for the other benefits. David explains:

> Often when I'm doing therapy with someone and they're talking about painful things from the past, I'll say, "So, what do you wish you'd been able to say back then? What do you wish you'd be able to do?" Usually, one of the problems with those situations is that people didn't have a voice. They didn't know what to say or they weren't allowed to say it. So they tell me, and it's a way of bringing some closure to that situation and giving themselves a voice. Well, in comedy, we take the same situations, but rather than just telling one person, we turn those situations into comedy material, then they get up in front of the theater of four hundred people and tell four hundred people who laugh and applaud. Of course, as soon as two or more people laugh together, there's this moment of instant recognition like, "Wow,

we are in this together." These people get this huge sense of validation and acceptance, which I think is a much greater sense of validation and acceptance than you get from just telling your therapist.

I had the opportunity to interview comedian Rick Shapiro, probably best known nationally for his work in the 2001 movie *Pootie Tang* or the TV series *Lucky Louie*, and his manager/wife Tracy. Anyone that knows Rick or his raw and honest stand-up will know that he has experienced more than his share of darkness. He suffers from PTSD and frequently battles depression. Rick's shows are a source of cathartic release for him, and in fact our interview was an emotionally charged session. As Tracy told me, it's as if the only therapy that works for him is the stage.

Again, the therapeutic benefits of performing have not been empirically tested, but I am certain that they are there. And there are many more than I've mentioned. Comedy gets you out of the house. I know plenty of comedians who, if they didn't have to perform, would spend all of their time on their couch, smoking weed and playing video games. With comedy, at least they leave the couch to smoke the weed. Seriously though, it is an incentive for activity. Doing shows can also expose us to new restaurants, parts of town, or entire cities that we have never explored so that sometimes we not only get off the couch, we get out of town. Finally, although when on stage stand-up is usually a solo performance, comedians do not work alone. Like any hobby, there is a great deal of potential for socializing in comedy. Some of my best friends are people I met through comedy. Because comedy attracts people from nearly every possible background, you will find yourself interacting with an incredibly diverse group of people. Having a good social network brings with it a whole host of other benefits. My financial advisor is a comedian I met on the road (although terribly unfunny), my real estate agent is a comedian, and I have received legal advice from comedians who practice law. As a psychologist, I have also counseled several comedians dealing with a variety of issues.

Stand-up comedy is not the only form of comedic performance art. Improvisation, or improv, as the cool kids call it, can also provide many of the same benefits. It might even be a more appealing option for those dealing with social anxiety or even just a mild case of awkward. Improv is performed by groups, and some might find this to be a safer environment than standing on stage alone with a microphone. There is also less emphasis on punch lines, so there is less potential to bomb. I am more familiar with the world of stand-up, so I spoke with author and clinical psychologist in training Daniel Wendler[95] about how he has used improv. Daniel has Asperger's syndrome and started doing improv theater in Austin (there's that city again!) after college. He found it not only fun, but also helpful in developing his social skills. He now incorporates improv techniques into his social skills coaching.

One of the basic rules of improv is the notion of "yes, and . . ." meaning that once a player says something or establishes a component of a scene, we accept that it has been established and build on it. "I am a giant termite looking for love" " . . . Okay Mr. Termite, welcome to The Log Roll, the hottest singles bar in the entire forest . . ." would be a decent "Yes, and . . ." response as opposed to "What? You look nothing like a termite; you are just a guy named daniel wearing a black T-shirt. I'm wearing a black T-shirt too! What are we, a gang?" (Okay look, I never said that all improv is comedy gold. Some of it can be hard to sit through, but when it works it works really well.) According to Daniel, learning the notion of "Yes, and . . ." lends itself well to the art of conversation. Because, as he put it:

> If you want to go deeper, you need to accept what the other person is bringing which is the "yes" and then you need to build on it which is the "and." The mistake that

a lot of people make is that I'll say, "Oh, it's a beautiful day today," and you'll just say, "Okay." You're not really building on it. You're not really doing anything with it. Whereas, a better response is something like, "Yeah. I wonder if we should go to the park." You take what I said and then you build on it and then that creates some opportunity for connection. That's a pretty insightful understanding of social interactions for a guy with Asperger's.

I am sure that improv can be very therapeutic, and can even help someone ease into stand-up comedy. As an up-and-coming performer myself, I am always looking for advice from my comedy heroes. I asked comedian, actor, and Emmy Award–winning writer Rick Overton if he had any advice for me and his answer was simple: "Study improv." He went on: "I say study some form of improv because whether you're serious about it or funny about it, improv just opens you up more and conquers more. If you can do improv, you can do comedy."

I would never recommend stand-up comedy as a career path. But if you have a decent sense of humor and are looking for a hobby that may benefit you in countless unforeseen ways, then yes I recommend giving it a shot. And if you happen to get famous, I'd love to open for you.

RECOGNIZE THE THERAPEUTIC VALUE
OF HUMOR AND LAUGHTER

A S WE NOW UNDERSTAND, HUMOR AND THE LAUGHTER IT inspires have a number of wonderful effects on our mental and physical health. Humor therapy is a term that can be used to describe any number of practices pertaining to the use of humor, but for most practitioners it refers to a very straightforward routine. This includes exposure to comedic material for at least thirty minutes a day.

As I mentioned, humor therapy has been practiced as early as the sixteenth century. Norman Cousins recommended a daily dose of laughter-inducing comedy to help suppress pain, and lots of research has confirmed various benefits to this prescription. I know many psychologists who regularly make humor recommendations to their patients, and organizations such as Cancer Treatment Centers of America are offering humor or laughter therapy as part of their integrative approach. There are also organizations of practitioners with an interest in humor. Most notably among these is the Association for Applied and Therapeutic Humor, an interdisciplinary organization founded in 1987 by nurse Alison Crane.[96] The humor therapy movement definitely seems to be growing.

I previously mentioned that when I am asked to provide recommendations for humor therapy, I find it important to get to know the patient's existing tastes a little first. We increase the likelihood of compliance if we make a recommendation that the person will enjoy or relate to. You may recall that for the woman who only watched the news on TV, I recommended that she start with *The Daily Show* as her introduction to humor therapy. I have made similar matches based on a patient's interests, personality, ethnic background, or health issues. Humor in general is helpful, but if we can relate it to the specific issues a patient faces, it can be especially helpful. For example, I have recommended the show *The Big Bang Theory* to people with Asperger's syndrome, as they would probably find many of the characters easy to relate to. I have recommended shows like *2 Broke Girls*, *Workaholics*, *Mike & Molly*, *Anger Management*, *Broad City*, *Community*, *Arrested Development*, and a whole bunch of others to people based on similar match-up potential.

TV shows sometimes have relatable characters, but on the other hand, comedians ARE relatable characters. Sometimes it is helpful to hear someone who knows what we are going through give us permission to laugh at our pain. Earlier I mentioned that Gabriel Iglesias has diabetes. Other than talking about how that inspired him to alter his life, I find that he does not discuss the illness very much in his stand-up. Therefore for most diabetics, I recommend they check out the work of Patrice O'Neal. O'Neal unfortunately passed in 2011, but he left us with some very hilarious takes on diabetes. Likewise, Maria Bamford talks about having obsessive compulsive disorder, among other mental illnesses. Josh Blue talks a lot about having cerebral palsy. Marc Maron regularly discusses his depression in his comedy performances as well as on his amazing podcast, *WTF*. Richard Pryor's various accounts of his drug addiction are hysterical and classics. In fact, lots of comedians discuss addiction. Lots of comedians, including the guy writing this sentence, discuss issues with weight management. A quick Google search is the only thing standing between you and a good recommendation for some good old doctor-prescribed humor therapy.

Just don't watch too much. Just like virtually everything else that is good for us, too much time spent watching comedy is likely to have a negative impact on our lives.

So it may seem to you that what I have just described is, in simpler terms, paying a therapist or other health-care practitioner to tell us to go watch TV. "Okay, watch an episode of *Family Guy* every day for a week and that will be $100 please." Holy crap, what a scam. Or it would be a scam if that was all there was to it. The humor viewing should be used in combination with other therapeutic activities. But I would suggest that, as my previous example of the woman who thought all pop culture was beneath her, there are some people out there who need to pay a therapist to tell them to watch TV. Some people become so overly focused on their career, their expectations of life, and trying to act like a so-called "grown-up" that they tend to cut humor from their lives. They tend to neglect their playful side, and are suffering as a result.

I don't know why, but I have always held onto my playful side as I've aged. So much so that when I was in my mid-thirties and I bought my first convertible, a friend said to me, "If it was anyone else, I would say you are having a midlife crisis; but that's just Brian being Brian." In fact, the very idea of a "midlife crisis" seems foreign to me. I am a little juvenile. I have an extensive toy collection, still like to play video games, and watch stupid movies full of explosions. My girlfriend, Sarah, matches me on this trait; she cannot pass up a playground without taking time out for a swing or spin around a merry-go-round. We've been accused of being immature, and perhaps we might be a little, but keep in mind that we are intelligent, educated professionals. You don't have to be serious to be an adult. Neither of us have any children, yet we enjoy occasional theme park visits.[97] On a recent trip to Disney World, Sarah, who is a therapist in case you've forgotten, commented to me how she thought so many of her patients would benefit from a simple visit to a park. Let loose and play once in a while—it's good for us at any age.

As I recount this next story, keep in mind that all of my work and pretty much everything I do, especially my talks, involves a fair amount of humor. I was giving a seminar on happiness, somewhere in the suburbs of Chicago. Like always, the people in the seminar had registered for it months in advance, responding to printed brochures. Because this was a lecture on happiness, humor was one of the subjects listed in the brochure. In the speaker's bio, also printed on the brochure, it stated that I am a stand-up comedian as well as a psychologist. A lot of people skip the speaker's bio, which is weird to me but I guess they assume that whoever is giving the lecture must be qualified to do so. In fact, I make a joke about this almost every day when I state that for those that did not read my bio, "I appreciate your blind faith in just handing over your money to any quack or charlatan that can print a nice brochure." The brochure also mentions that "audiences appreciate his use of humor." In other words, there should be no way an informed person could come to my seminar and not expect a little humor.

Apparently, however, there is a way and on this particular day a woman was determined to prove it to me. The setting was a crowded conference room at a DoubleTree Hotel and she sat right there in the center of the three hundred or so people in attendance. I went through my usual morning material, which includes some introductory jokes and other stories like I have shared with you here. A couple of hours in, and about fifteen minutes before our lunch break, she raised her hand. I asked her if she could hold her question until the break and continued with my lecture. She couldn't. She then did something that I have never seen anyone do before; at a point where the rest of the audience was enjoying themselves and laughing, she stood up in the middle of the room and shouted her comment to me. She said, "We get it, you are funny. But I didn't come here to laugh, I came to learn about happiness!"

Without a hint of irony, she literally asked me to stop telling jokes so that she could learn how to be happy.

As a comedian, I am used to hecklers, people who interrupt the show or distract from it, but not as a public speaker. As a speaker, most of the ridiculousness I face comes from people asking me to alter the temperature of the room ("It's too hot!" "It's too cold!"), as if the control of the air conditioning is in my pocket, or from those who apparently think I should stop my lecture to serve beverages ("We need more coffee!"). Those types of stupid requests are easy to deal with, but this was the first time I had an interruption anything remotely like this and it caught me off guard. The rest of my audience had been enjoying themselves and learning, but now there was this dark can of stink that had been opened up right in the center of the room. I had to deal with it. I said, "Point taken. Now, you should know that when I made reference to people that did not read the brochure correctly, I was NOT asking you to self-identify. But thank you for giving me an example of how people can lose their appreciation of humor and fun."

Some people can be such a buzzkill.

I did what I could to keep my momentum going, but the comment had really thrown a monkey wrench into my mojo. I finished what I had to say and broke for lunch to shake it off. By the way, during the lunch break many people came up to me to make sure I knew that the rest of the audience did not share her opinion, and they were enjoying the seminar. Unfortunately, the woman that stood did not return to the class after lunch. Too bad, because it seemed as if she really could have used it.

There is no need for us to take life that seriously. Author Brian Sutton-Smith, who coincidentally was also a professor at my alma mater, Bowling Green State University, for a while (because only the finest scholars are willing to brave northwest Ohio winters), cautioned against this. He said, "The opposite of play is not a present reality, or work; it is vacillation; or worse, it is depression." We don't have to give up play as we age, and we don't have to give up humor to take a subject seriously.

Watching comedic movies or TV shows or even listening to comedians can potentially help us learn to rediscover our lost sense of play, but we can also learn to take life less seriously without regular exposure to humor. I spoke with the director of Mind-Body Medicine at Cancer Treatment Centers of America, who originally began their humor therapy program more than eleven years ago at the request of a patient. At first, they inspired laughter in their humor groups by watching movies and other videos. Eventually, they found it really difficult to find stimuli that were suitable for all group members. Everyone has different sensibilities. Many comedies involve profanity, or jokes about race, sex, religion, or politics and as I've mentioned, anything (even cookie dough) is potentially offensive to someone. So given the difficulty in finding stimuli, the CTCA moved toward practices like Dr. Kataria's laughter yoga. Through these practices, they could generate the laughter experience without depending on stimulus materials or jokes.

Laughter groups are now led at all CTCA centers. Patients participating in these laughter groups often report benefits; however, there is no formal research demonstrating the effects. This is generally true across most settings that use humor therapeutically. Although research does show that humor is beneficial, research on the use of humor therapy has been mixed. Some studies do show a positive impact, whereas other studies fail to show any effect. As far as I am aware, there are no studies that show a negative effect of adding comedy to a patient's diet. My personal thought on all of this is that because it is very inexpensive and doesn't require a tremendous time commitment and may help (but if it does not, then at worst, it will do nothing), if it is something you are considering, go ahead and do it. You have very little to lose, and it might help in the long run. Besides, it's not like you aren't already planning on watching a little TV, so watch some damn comedy.

As I mentioned in the chapter on pain, dentists often use humor to help patients relax and establish a rapport. The same can be true for any medical professional. Humor helps develop a better doctor-patient relationship and ensure greater adherence to treatments. Imagine how much more likely a person would be to comply with their doctor's orders if they actually liked coming into the office. That's such a nutty concept it almost seems unheard of. That is kind of the idea behind Dr. Hunter "Patch" Adams's Gesundheit! Institute.[98]

The institute, whose early years are portrayed in the movie *Patch Adams*, represents a novel approach to health care. The caretakers operate on several principles including the belief that health care should be provided for free (which sounds very appealing to this guy) and that patients should be treated as friends. If you have seen the movie, which from what I understand is not an accurate characterization of Patch but more of Robin Williams, then it shouldn't surprise you that they also believe care should be infused with fun and play. They even offer classes on clowning.

Humor can be an important component in interpersonal therapy, and might even make some of our existing methods more effective. For example, psychologist Larry Ventis uses humor exercises to help patients with irrational fears, called phobias, become desensitized to their triggers. Systematic desensitization is a traditional method of regularly exposing a patient to the stimulus that provokes their fear response at different levels of intensity until they are able to remain calm. For example, suppose we have a fear of heights we are trying to overcome. Well, for one session we may start out by standing on a chair, not a very intense height, until we get used to it. Then maybe we move on to a tabletop, until we get used to it. Then, we take out that ladder we borrowed from our neighbor so many years ago and forgot to return. . . . Eventually we take an elevator to Skydeck on the 103rd floor of the Willis Tower in Chicago and look STRAIGHT DOWN as we are suspended against all logic and reason over the busy streets of downtown Chicago in one of the most ridiculous photo opportunities ever. You get the idea. Ventis and his associates[99] (2001) have shown that humor exercises can help reduce the anxiety we feel when we are exposed to the object of our fears. By learning to laugh at them, the traditional method is made even more effective.

PRACTICE LAUGHTER,
AND LAUGH OFTEN

I LOVE THE CITY OF NEW ORLEANS. I ABSOLUTELY LOVE IT, DAYTIME muggings and all. I first went there by accident on a cross-country trip with my brother Jon. We were in college in Austin at the time and drove to New York for a summer visit with the family. At the end of the summer, we took off, first heading south on I-95, planning on catching I-40 to head southwest. Because I was older and it was my damn car, I did most of the driving, but somewhere through the Carolinas on our trip home, I let Jon take the wheel so I could get some sleep. When I woke up, we were approaching Jacksonville, Florida, as he had missed our turn off. Now, we had to reroute by taking I-10 all the way across the South, and as we entered into New Orleans, I insisted we stop and check it out.

As soon as I set foot in that city, somewhere off Canal Street, it was love at first sight. At that point of my life I hadn't explored many places on my own, but I was impressed with New Orleans. The architecture is amazing and unique, the history is rich, and the culture definitely agrees with me. There is a Cajun expression, *laissez les bons temps rouler*, which means "let the good times roll," that has pretty much become a motto for the city. Good times can be anything. Sure, New Orleanders are known for their hard-partying holidays, but they also love their food and music. They regularly dance in the streets and laughter is abundant. I spent maybe two hours in the Crescent City on that accidental detour. However, it was long enough to make such an impression that I decided then that regardless of what I did for the rest of my life, I knew that I had to spend some of it in New Orleans. This desire was behind my application to graduate school and I moved there the following year, right after college.

I've mentioned my grandfather a few times. Before he passed we had a discussion about our genealogy. I learned that, coincidentally, the man who carried my version of the King name from Europe and eventually gave it to the man who gave it to my father (not Martin Luther King, despite the resemblance) entered this country through the port of New Orleans. Moving there, I felt as if my family had come full circle.

New Orleans, as you probably know, was devastated in 2005 by Hurricane Katrina. I remember at the time, I was living in Philadelphia, but my heart was with my favorite city. When the National Guard started letting people back into the city, I went down for a visit and explored some of the most affected neighborhoods with a few friends. We saw firsthand some of the craziest destruction ever. Houses lifted from their foundations and smashed into each other, debris piles comprised of building materials and ruined household possessions, major appliances rendered useless, and unidentified cars and boats were scattered everywhere. One friend of mine was looking for his car, which had been carried away by floodwaters, and eventually found it on top of a house several blocks away. He had lost everything to that storm, but at the sight of seeing his car upside-down on top of a damn house, he burst into uproarious laughter. I joined him and all we could do was stand there and laugh away the tears. The devastation was horrible, laughter was our defense.

Before Katrina, New Orleans was no stranger to tragedy and hardship. It is built below sea level, and although 2005 was a particularly nasty storm, the city regularly gets in the way of hurricanes. It has also had a struggling economy, high crime rates, and has dealt with political corruption for decades. Now I am no sociologist, but it seems that despite all the negativity it has experienced, the city still seems to function and its citizens love the hell out of it. I have a theory that the fun-loving, "let the good times roll" attitude of New Orleans is one of the main reasons its population can remain so resilient in the face of so much negativity. The ability to laugh it off is as much a part of the culture there as grits and gumbo.

If there was ever a place that recognized the need to not take life so seriously, it was New Orleans. The last time I was there, I saw a sign on the street that said simply, "Embrace Happiness; Laugh More; Worry Less!" and I thought to myself, *Is there any question why this is my favorite city?* The people of New Orleans know how to live because they know how to laugh.

I always felt a connection with New Orleans, and then for some reason I went and moved to Bowling Green, Ohio, famous as the home of the National Tractor Pulling Championships (yes, that is a thing) and where they didn't get my sarcasm ... I don't always make the best decisions.[100]

Laughter is a behavior. Generally, it is our response to humor but a lot of laughter is expressed in the absence of humor. We laugh to punctuate conversation. For example, a little "ha ha" at the end of a spoken sentence can serve the same purpose as an "LOL" at the end of a text. That form of laughter may benefit our communication, but it is the longer-lasting hearty laughter that is believed to be responsible for most of the benefits I have listed in this book.

Like the sign in New Orleans read, we should worry less and laugh more. We should laugh often, it's good for us. Unfortunately, not all of us laugh as much as others. Some people for various reasons do not laugh. We may have an innate potential to laugh, but we have to learn how and when to use it. Like any behavior, laughter takes practice. We did not learn how to walk on our first try. We didn't learn how to speak with our first drooly mumblings, and we don't learn to laugh without the right socialization and practice. You know that expression "practice makes perfect"? Well, technically speaking, practice actually makes habits. Every habit we have ever acquired, whether it is biting our nails, eating high-caloric foods, wiping our hands on our pants, or leaving the seat up on the toilet (Can you tell my girlfriend helped with this list?) was learned by your brain by simple repetition, i.e., practice.

Almost all of our brains know how to produce a laugh, but for some of us what is missing is the brain's ability to automatically respond to a stimulus with laughter, in other words the habit. As I've mentioned before, there are many reasons that we may not laugh, but unless it is due to significant brain damage, it really doesn't matter why. We just need to start practicing.

I've mentioned repeatedly that modifying our existing habits is extremely difficult. If we have already developed a set of responses that our brain has automated, it is going to take a tremendous amount of cognitive awareness and effort to redirect our behavior into something else. For example, instead of exercising I have the habit of sitting at my computer too much. Sure, I work from my computer (I'm doing it now actually) but I also use it to play games, watch videos, and listen to music. That's a lot of computer time, and I would still rationalize that as productive use but the thing is, and I am sure I am not alone here, the majority of my time starring at the computer screen is spent mindlessly surfing the Internet and mostly Facebook. I would classify that as a bad habit; it is at least a physically inactive one.

New habits are really difficult to acquire, just ask anyone who has ever attempted to start exercising later in life. Actually, you could just ask me. I hate exercising, but I know I need to do more of it if I'd like to achieve my goal of having abs so washboard-like that my washing machine gets jealous and, of course, my long-term goal of living forever. My friend Dave works out six hours a day, which is just ridiculous, but I admire his discipline because I have a hard time getting in six minutes. Occasionally I'll pick up my hand weights or use the treadmill at the gym, but I don't do it often enough to develop the habit.

Practice makes habits. Thankfully, laughter is a relatively easy behavior to practice. It is easier than lifting weights and jogging in place. If you are not a natural laugher, that behavior can be brought out through help and special exercises. Earlier I discussed a movement called laughter yoga, founded by Dr. Madan Kataria. Laughter yoga is based on the belief that voluntary laughter is just as beneficial as, as I put it, organic laughter inspired by humor. It consists of breathing exercises designed to mimic organic laughter that are practiced in a group setting. The exercises turn into authentic, gut busting laughter fairly quickly. There is not a ton of empirical research available on the subject, but it does appear that laughter yoga practitioners experience some real benefits. For example, in some preliminary studies, it has been shown to reduce depression in elderly women,[101] improve heart rate,[102] and increase general health.[103] Similar practices, with names like "laughter meditation," "laughercize," and "laughter base jumping" (okay, I made that one up) have also become popular.

Anecdotally, laughter yoga does seem to provide people with the same benefits as organic laughter, and I see very few potentially negative side effects to the practice. Therefore I do like to recommend it to people who may otherwise have trouble laughing. It can be practiced on a regular basis, and as a result can help someone acquire the habit of laughter. In fact, many people have reported that laughter yoga, or something similar, has helped them to "reconnect" or "rediscover" their laugh. So, practice.

It's just like what the New Orleans jazz musician said when a kid asked him how you get to Preservation Hall: "Practice." ·

FINAL THOUGHTS

WHEN I WAS IN COLLEGE IN AUSTIN, I WAS A HUGE FAN OF *THE Jerry Springer Show*. Unlike other talk shows, it came on late at night and was absolutely absurd. Jerry would host discussions on outlandish topics, usually involving some sort of scandalous, and possibly incestuous,[104] infidelity, and almost every episode would spontaneously break out in fights or nudity or both. Unlike other shows of that genre, Jerry didn't seem to be interested in helping his guests, just letting them pummel each other or rip each other's wigs off on stage in Chicago for a few minutes. It was base-level entertainment, before such shows became the norm, and my friends and I ate it up.

Apparently it is still on the air in some form, but I remember the early years of the show.

My favorite thing about the show was Jerry himself. Despite the train wreck playing out on his stage every day, he was a very intelligent and well-educated man. In fact, I sometimes saw parallels between his life and my life. His family settled in New York, and I was born there. He went to school in New Orleans, and so did I. He earned a graduate degree, as did I, and his professional career has taken some very unexpected turns, as has mine. Ultimately he landed on TV, and here I am, a comedian with Hollywood aspirations. At the end of each show, he would let his education shine through with a very thoughtful and articulate "final thought." Here, then, is mine.

Over the course of these pages I have discussed how wonderfully awesome laughter is. It is one of my favorite things to do, to laugh. It brings people together and contributes to our quality of our life. It has a multitude of health benefits. However, as beneficial as laughter can be, it is important to keep in mind that it is not a cure for anything. It helps our immune system but is not a substitute for any treatment, so let's not toss aside our medication in favor of a few good daily laughs. It can help reduce or ease our pain, but it is also not a magic pill. It can help reduce our stress, and suppress some of the destructive thoughts that provoke anxiety and depression, but laughter alone cannot prevent or eliminate disorder.

Improving our mental and physical health is a long-term pursuit that might involve us making some really difficult changes to our lives, but laughter can ease the process. It can be the lubricant that helps the machine function. At the bare minimum, laughter makes us happy. Certainly nobody needs to be happy, plenty of miserable people live productive lives, but as I often remind people, we only have one life to live, and we don't know how much of it we have left. Our behavioral choices impact the quality of that life. We can decide that things are fine the way they are, we can decide (perhaps not consciously) to make them worse, or we can decide that we'd like to make them better. The choice is ours. As for me, I prefer happiness over sadness and health over illness, so I choose to take life less seriously. Hopefully, you will join me in this and let's all add a little more laughter to our lives.

As Jerry Springer would say as he'd sign off, "Take care of yourself, and each other." I would also like to add, "Please, whatever you do, try to stay off the *Jerry Springer Show*." Or whatever the modern equivalent is.

Hopefully this book has been informative and maybe it even made you laugh at parts. If I've done nothing else but inspire a few laughs, I consider that a job well done. Laughter is healthy and good for you. As I write this last section, I am about to embark on another cross-country speaking tour. Maybe our paths will cross, and maybe we will get to share some laughter in person.[105]

I love to travel, and I do so a lot. In this section, I have recently written a lot about my love of the city of New Orleans. It is one of my happy places. As of this moment, my girlfriend, Sarah, has never been, and tomorrow I am taking her there for a few days of vacation before I start my tour. Sarah, being the great motivator she is, won't let me go until I finish this passage. Truthfully, the only thing between me and a trip to New Orleans in the morning is coming up for an end to this damn book. Well, I am a simple guy and I started off with a simple "Hi" so it seems only fitting that I end in turn.

Bye![106]

And thanks for reading.

APPENDIX I: HUMOR TESTIMONIALS

THROUGHOUT THIS BOOK I HAVE SHARED A LOT ABOUT MYSELF, and most of it was even true! My sense of humor and laughter has helped me a lot over the course of my life. It helped me make friends when I frequently moved during my military upbringing, it helped me talk my way out of getting my ass kicked, it helped me get out of a couple of tickets, and it helped me score a cute chick once in a while, the often-mentioned Sarah most definitely included. It has probably helped me postpone health complications due to some of my other lifestyle choices (my mother is still amazed that my blood sugar is never high). Professionally, my sense of humor made me a better professor, something noted in almost all of my teaching evaluations. More than my training in psychology, my experience as a comedian has made me a better public speaker. Ultimately, my sense of humor led to my opportunity to write this book for you.

This book is not about me; it is about humor. I like to use personal examples from my own life and experiences as much as I can because the topics I discuss are personally relevant. Humor has impacted my life in a major way, but it has impacted the lives of countless others as well. As a touring speaker and comedian I encounter people all the time whose lives have been touched by humor. They range all over the spectrum from professional comedians to dentists, comedy fans to teachers. I have asked a few of those people to share their stories with us, and have included many quotes throughout.

Here are some additional testimonials to the value of humor.

BRYAN, COMEDY FAN (GAINESVILLE, FL)

I STARTED SUFFERING FROM DEPRESSION AROUND FOURTEEN. I slowly started realizing that I wasn't cut out for a lot of things at school academically, and I was losing a lot of friends and being bullied a lot. I figured I'd rather be dead than wake up and go to school every morning just to hate myself and who I am.

My first attempt was to jump off the roof of my friend's house (I was dog sitting for them). I got drunk and decided to just jump. Thankfully, booze gets your body relaxed and makes impacts like that less effective. I walked away with a minor concussion. My second time, I tried to hang myself, but the belt broke. The third time I tried hanging myself again, but the rope broke. After that third time, I felt even more disappointed with myself.

I remember my dad introduced me to George Carlin, and I was always laughing so hard when listening to his rants. One rant he did was on suicide, and I couldn't help but laugh. From then on, I viewed my failures as a joke as to how I'm so much of one that I can't even kill myself properly.

But Carlin saved me. I then got to hear about other comedians. Brian Regan, Seinfeld, Jim Gaffigan, etc. Whenever I felt depressed again, I decided to just hit the Internet or Comedy Central and just laugh until I forgot about my life for a while.

Comedy saved me. I'm on medicine now, and that keeps me from feeling depressed, but I know I'll always have comedy to hold me up.

GLORIA RHYNE, CANCER RESEARCHER (PORTLAND, OR)

WHEN I WAS FOURTEEN, I WAS DIAGNOSED WITH A PARTICU-larly difficult form of cancer called osteosarcoma. I spent what should have been my freshman year of high school in California with my mother receiving treatment for something that statistically should have killed me or greatly maimed me at the very least. However, being the punk I always was, I found ways in which to entertain myself: by "accidentally" dropping my crutches for the cute physical therapist to pick up, by luring the young fellows into my hospital room to watch *Monty Python's Flying Circus* with me, or just by telling people to rub my bald head for good luck. To this day, when I tell someone of my experiences, I always state that my biggest regret is that I never fully developed my superpowers after chemotherapy.

JOHN AGER, COMEDIAN (RENO, NV)

W E KIDS WERE RAISED IN A NOT-SO-FRIENDLY HOUSEHOLD, SO we used humor to get through the rough times. Our mother and father knew this, and were sometimes the butt of our jokes. Outsiders did not always get it, which helped us to laugh more often at their reactions than the actual situation. Mom had been diagnosed with a brain tumor, which the doctors said would give her two to four months to live. We were all shocked, and told mom so, as we had no idea she even had a brain. This gave way to her calling us "you little bastards," a dig on her actually, and all of us laughing.

Mom went through a series of therapeutic tests to see if she could stay in her apartment in the senior living center. The tests showed she could be a danger and that is what the panel of therapists and doctors told her and I later that day. When I asked her what she wanted to do, she said she just wanted to go home and finish her days in her apartment.

I told her in front of the panel that I would love that also; however, we did not have enough money to get her the full-time care she would need. In fact, I told her, if you don't die within the two-to-four-month time frame the doctor gave you, we won't be able to afford you and you will be out in the street.

She looked at me, while the panel of doctors and therapists stared with their mouths wide open, shocked looks all over their faces, and said, "You little bastard, I will come back and haunt you," and we both burst out laughing. This of course allowed the panel to relax and see how crazy we were, and the tensions just flowed away.

KATRINA BROWN, COMEDIAN
(CANTON, OH)

HAVING BEEN BORN WITH A RARE BIRTH DEFECT EVEN DOCTORS have never understood, I spent the majority of my life feeling like an outcast. That was my reality, though, so it never felt exceptionally bad, or good. It just is what it is. I noticed young that people will point and laugh at things they don't understand. I learned young you [have to] laugh with them or spend a lot of time hating yourself for things you have no control over. When I make jokes about the metal plate in my head, it morphs into whatever "imperfection" an audience member has in their own life. While being born without the majority of my skull might shock some people to hear, telling them about finding the humor in it eases their minds about their own issues.

We live in a world now where everyone wants to stop bullies. We can't. They're everywhere. What we CAN do is teach people to use humor as a shield against it. I was picked on, teased, it was assumed I was "retarded" for the protective helmet I wore until age five. A stranger once smacked my helmet off my head at the mall, yelling at my mom for putting such an ugly thing on a child's head, not knowing her act of ignorance could've killed me. Serious stuff . . . that's still funny. I wore a HELMET the first five formative years of my life in an age NO KID wore helmets. I milked that shit for extra candy at Easter. I played the "wounded child" card with grandparents knowing

FULL AND WELL what I was doing. And now, with a plate in my head? I get to mess with security guards at metal detectors.

To take the worst of who we might be, and be able to laugh about it is utterly freeing. You take this unfortunate event and decide YOU have control over it. YOU are going to make it what it is going to be and how it affects your life. At forty-four, I still suffer massive headaches from the plate. In hotter climates too much sun exposure can lay me out for days. When you're a road comic and get booked in Florida in June? You have hurdles to leap. But? YOU control this thing. It doesn't control you. You take what is "broken" and you decorate it until it's beautiful. I've had people with various disabilities and oddities stop me after shows thanking me for taking something personal and finding the humor. THAT makes it all worth it.

MACK STEVE, COMEDIAN AND WRITER (LAS VEGAS, NV)

Well I'll get right to it and first lay out the adversity in my life. First I was the last child of four in a Mormon family and an accident during the hard financial times in the eighties, so unpopular in the family, especially since I was liked more than them by everyone else. I wrecked my body in athletics. My lower back has become the bane of my existence and what lead to my becoming heavily addicted on pain pills after four back surgeries. After the RX pills drugs addiction had taken everything, my life, limb, job, and my soul, I decided to get off pills.

After deciding pain pills made life as interesting as Iowa and they made the idea of falling into traffic ever more appealing, I got off. It's not that my pain has disappeared; I have gone through nineteen surgeries and broken countless bones, or my other issues [have] gone away. I no longer use any pills whatsoever. What I use is humor, cannabis, meditation, and diet as much as I can. Somehow I'm able to take the pain out of a situation with comedy. For comedy I dropped twenty-five pounds, got a leg up, a foot in the door . . . made sure I couldn't be run off stage with one few swoop. What it's done as well is made me slow down to see the beauty in life. And seeing how my webpage is www.gimppimp420.com, pimp living life to a higher standard.

SANDI SELVI, COMEDIAN AND WRITER
(SANTA CRUZ, CA)

MY STORY BEGAN ONE NIGHT WHEN I WAS GOING THROUGH MY stem cell transplant. March 15, 2000. They had to put a port in my chest before they could harvest my stem cells. When I woke up in recovery I couldn't breathe right. When I told the nurse, she thought I was nervous so she gave me an Ativan to calm my nerves. Twenty minutes later I still couldn't breathe. She gave me two more Ativan. After that I couldn't have cared less that I couldn't breathe. The Ativan was working, no anxiety at all. So they sent me home. I was so tired when I got home my husband and I went to our separate rooms (I was isolated) to sleep. My husband fell asleep in two seconds; he was exhausted and needed sleep. The second I laid down I realized I could not breathe at all. So, I propped myself up with all the pillows in the house, and then tried to find something to keep myself awake until my appointment at 8:00 in the morning. I had bought some comedy tapes from Costco. I played the first tape, Brian Regan. I laughed so hard that I actually got air into my lungs. My husband finally came in to see what I was doing, took one look at me and ran down the hall. He started the car, ran back, picked me up, and said, "We have to get to the hospital. Your lips are blue." I said right then and there, "If I live through this, I'm going to be a comedian." He rushed me to the hospital. Turns out they had punctured my lung by accident during the port insertion. I believe if I had not listened to those comedy tapes that night, I would have died.

TIMOTHY, COMEDY FAN
(PITTSBURGH, PA)

WHEN I FIRST FOUND OUT THAT ROBIN WILLIAMS HAD COMMITted suicide, I must admit that I was not surprised. Like the old joke of the great clown Pagliacci, it is true that only the saddest of minds can truly understand how important and magical comedy is. It was a shame that he succumbed to the sadness, but perhaps he didn't have enough in him left to laugh again.

That is why I want to emphasize the importance of laughter. I have suffered from depression throughout the majority of my life. And as I entered adolescence, it became exponentially worse. I hated almost all of my life. But my friends and I would get together for a half hour in the cafeteria and make each other not just laugh, which would have been enough, but completely lose our minds in the hysteria of the horrible world we lived in. I survived my day for the jaw-aching laughter that we produced within each other. And as I got older, as life became sadder, even into my twenties, I began to truly realize that in my saddest moments, in my darkest corners, I would occasionally still find myself laughing at bits about airplanes and bad relationships. To be able to create laughter in the midst of misery is an extremely unique and vital skill. When I felt my heart broken, a sad clown also knew what that felt like. When I felt worthless, someone famous for making people laugh had been there. And in my most self-abusive, a real comedian would be able to look right at me and laugh, saying, "Oh, man, I don't want to remember all the times I've felt that sting!" And this unity may very well have saved my life.

So to all of the greats who have worked so hard to make something out of their sadness (to name so very few, George Carlin, Sarah Silverman, Brian Regan, Patton Oswalt, Ron White, Dave Chappelle, Larry David, Brendon Small, and my personal hero and savior, the great Louis C. K.), I have a very important message for you all: You are not just entertainers. I actively believe you are saving people's lives. And from the bottom of my healing heart, I thank you.

TREVOR HARRIS, COMEDIAN
(LOS ANGELES, CA)

I WANT TO START BY SAYING I NEVER PLANNED FOR THIS WHOLE comedy/acting aspiration. I was just a loner who grew up with cerebral palsy in Dallas who was in and out of hospitals due to double digit brain surgeries. Then, one day, I was playing basketball at thirteen after moving to LA, broke a couple ribs, and when I got home, pain started in my stomach, and then, went to my balls, legs, ass, feet, and up to my head before shooting everywhere while I was in the ER. My mom and I didn't know what was going on. And I got transferred to a terrible place in Long Beach where the doctors were so bad I was threatened with a psych ward stay for a while (which caused me severe PTSD that's now being cleaned up in therapy) because everyone thought I was faking my pain. When I told my mom, she got me out of that place and took care of me for a couple years while I couldn't move an inch without screaming and crying in pain until I got into UCLA where they said I suffered from my nervous system completely crashing due to the broken ribs being the last consistent trauma my body could take before snapping. I was put on ketamine (aka Special K), a horse tranquilizer, that made me able to stand up. But not completely heal. The dreaded day came where I was ordered by the only doc I trusted to be taken off all the magical drugs keeping my pain bearable. I was in a wheelchair and she said,

"We're going to take you off all these meds because we believe it's prolonging your illness. California hasn't been good to you, has it? You need to move back to Texas to get better. You need to laugh more. Try stand-up for your pain." I was surprised she suggested that until I remembered watching a roast at fourteen after being in pain 24/7 for a whole year and during that whole hour, I was pain free. I felt like I was floating.

So, tried comedy just for pain therapy. And every time I got up there, I was in no burning, stabbing, squeezing, shooting pain, and without a single migraine. When I decided to take stand-up seriously, things started to happen. I was offered a short film attempting to inspire those who are where I was medically that they'll have normal lives again. I was also offered a documentary recently, based on the same premise, that will start sometime in the near future. Things have been going up ever since I started this journey. And that's my advice to anybody out there in pain. Find your passion, stick with it, and you'll find people who don't think you're a lost cause because of your chronic illness.

WILLIE MAC, COMEDIAN
(SACRAMENTO, CA)

I STARTED IN COMEDY WHEN I DECIDED TO END MY TWENTY-THREE year marriage.

I was a shy and insecure kid, too many reasons to list. At fifteen I decided that this wasn't going to get me anywhere in life, so I got a job at the snack stand at the San Rafael bus terminal and made myself get used to talking to strangers. That led to retail, retail management, and later [I] started training to be a motivational trainer.

One Sunday afternoon in 2008, at the age of fourty-nine, I didn't want to go home. Wife was on the warpath and drunk/ hungover, and a real problem. After so many years in a dysfunctional, and some say a co-dependent, marriage, I was whipped. I meet two guys at a bar (Jason Resler and Kul Black) and when I found out they were working comedians, I mentioned that I had thought about doing it. That is when Jason jumped in my face and challenged me. Double dog dared me! This caused me to reflect on why I had not done so. I had written some material a few years earlier. The old feeling of fear and insecurity were holding me back, just like it had kept me in a dysfunctional marriage for so long. So it took me two months to sack up and go to Laughs Unlimited for open mic.

When I was done I had to ask myself why I had waited so long. I found something that made me happy, it gave me the opportunity to meet new friends, make people laugh, and I found a new sense of achievement. I had found happiness after many years of a life with more crap than I thought I deserved. . . . The performance of comedy has renewed confidence and excitement for life. After almost seven years I am still doing stand-up. I have a beautiful partner for life that stalked me from the beginning and a case of self-inflicted schizophrenia since I use a stage name and have moved to a small town.

APPENDIX II: EXPRESSING GRATITUDE

In its short time on this planet, positive psychology has uncovered some very interesting phenomena regarding happiness. One very robust finding is that our happiness is related to how often we express our gratitude to others. This is not a book about happiness, but as a generally happy person I like to express my gratitude often, and whenever appropriate.

Unless you were only pretending to read this book, you probably noticed that I have included quotes from several comedians throughout. And if you were pretending to read it, what the hell for? And why then are you taking the time to read this part? Anyway, in preparing this work I reached out to a lot of comedians in my social networks. Some I have worked with and know personally, others I have connected with via our common interests, and a lot of them submitted jokes and stories, and answered my questions. I may not have been able to use material from everyone on this list, but I would like to formally thank these individuals for their time. I would also like to encourage you to check them out online or at a show near you.

John Ager, Reno

Greg Asdourian, California (www.gregasdourian.com)

Katrina Brown, Midwest (www.katrinabrowncomedy.com)

Kendra Cunningham, New York
    (www.kendracunningham.com)

Vance Degeneres, National

Dave DeLuca, California (www.daveishere.com)

Ariel Elias, New York

Geoff Grant, Las Vegas

Trevor Harris, Los Angeles (www.youtube.com/comedytrevor)

Zorba Hughes, San Francisco

Ryan Kasmier, San Francisco

Willie Mac, Sacramento

Rudy Martinez, National (www.rudyisfunny.com)

David Mencarelli, Reno

Rick Overton, National (www.realrickoverton.com)

Joe Padilla, California (www.dammitjoe.com)

Dom Paré, National

Chelcie Rice, Atlanta (www.chelcierice.net)

Abby Rotstein, Sacramento

Sandi Selvi, California (www.sandiselvi.com)

Rick Shapiro, National (www.rickshapiro.tv)

Stan Silliman, Midwest (www.stansilliman.com)

Yakov Smirnoff, National (www.yakov.com)

Mack Steve, Las Vegas (www.gimppimp420.com)

Jason Ward, Boise

ENDNOTES

1   Run by comedian Kurtis Matthews, www.sfcomedycollege.com.
2   Colton, C. W. and Manderscheid, R. W. (2006). Congruencies in Increased Mortality Rates, Years of Potential Life Lost, and Causes of Death Among Public Mental Health Clients in Eight States. *Preventing chronic disease*, 3(2).
3   Hara Estroff Marano, "Depression Lowers Productivity," *Psychology Today*, July 1, 2003, https://www.psychologytoday.com/articles/200307/depression-lowers-productivity.
4   S.F. Maier, M. Laudenslager, and S.M. Ryan, "Stressor Controllability, Immune Function, and Endogenous Opiates," in Affect, Conditioning, and Cognition, 203–10.
5   See, e.g., K.Ranga Rama Krishnan, Randal D. France, Susan Pelton, Una D. McCann, Jonathan Davidson, and Bruno J. Urban; Chronic pain and depression. I. Classification of depression in chronic low back pain patients, Pain Volume 22, Issue 3, July 1985, 279–87
6   Cannon, W. B. (1927). The James-Lange theory of emotions: A critical examination and an alternative theory. *The American Journal of Psychology*, 106–24.
7   Strack, F., Martin, L. L., and Stepper, S. (1988). "Inhibiting and facilitating conditions of the human smile: a nonobtrusive test of the facial feedback hypothesis." *Journal of Personality and Social Psychology*, 54(5), 768.
8   Lewis, M. B. and Bowler, P. J. (2009). "Botulinum toxin cosmetic therapy correlates with a more positive mood." *Journal of Cosmetic Dermatology*, 8(1), 24–26.
9   Lefcourt, H. M., Davidson, K., Prkachin, K. M., and Mills, D. E. (1997). "Humor as a stress moderator in the prediction of blood

pressure obtained during five stressful tasks." *Journal of Research in Personality*, 31(4), 523–42.

10   Surwit, R. S. and Schneider, M. S. (1993). "Role of stress in the etiology and treatment of diabetes mellitus." *Psychosomatic Medicine*, 55(4), 380–93.

11   Bennett, M. P. and Lengacher, C. (2009). "Humor and laughter may influence health IV. humor and immune function." *Evidence-Based Complementary and Alternative Medicine*, 6(2), 159–64.

12   Skevington, S. M. and White, A. (1998). "Is laughter the best medicine?" *Psychology and Health*, 13(1), 157–69.

13   Levitt, E. B. "University of Maryland School of Medicine study shows laughter helps blood vessels function better." Publication by University of Maryland Medical Center. (March 9, 2009) [cited Oct 25, 2011], http://umm.edu/news-and-events/news-releases/2005/school-of-medicine-study-shows-laughter-helps-blood-vessels-function-better.

      See also Hayashi, K., Hayashi, T., Iwanaga, S., Kawai, K., Ishii, H., Shoji, S. I., and Murakami, K. (2003). "Laughter lowered the increase in postprandial blood glucose." *Diabetes Care*, *26*(5), 1651–2.

14   Social laughter . . . which, in turn cites:
Cogan, R., Cogan, D., Waltz, W., and McCue, M. (1987). "Effects of laughter and relaxation on discomfort thresholds." *Journal of behavioral medicine*, *10*(2), 139–44.
Zillmann, D., Rockwell, S., Schweitzer, K., and Sundar, S. S. (1993). "Does humor facilitate coping with physical discomfort?" *Motivation and Emotion*, *17*(1), 1–21.
Rotton, J., and Shats, M. (1996). Effects of state humor, expectancies, and choice on postsurgical mood and self-medication: A field experiment. *Journal of Applied Social Psychology*, *26*, 1775–94.

15   Anxiety disorders are the most common mental illness in the US, affecting forty million adults in the United States age eighteen and older, or eighteen percent of the population. Source: National Institute of Mental Health, http://www.nimh.nih.gov/health/statistics/prevalence/any-anxiety-disorder-among-adults.shtml.

16   World Health Organization (WHO) report quoted in BBC-Online January 9, 2001; see also "Global depression statistics" July 26, 2011, BioMed Central, available at http://www.sciencedaily.com/releases/2011/07/110725202240.htm.

17   Gray, P. (2011). "The Decline of Play and the Rise of Psychopathology in Children and Adolescents." *American Journal of Play*, *3*(4), 443–63.

18 Determining the effectiveness of a treatment is a complicated issue. My statement is based not just on the efficacy of medical treatments, but also on the success rates of therapeutic interventions. The measure of success varies from study to study. If we measure success as overcoming depressive symptoms without relapse or dependency on medication then our existing methods are ridiculously ineffective. One in three patients relapse within a year, and for most people depression is a lifelong disorder. Contributing to this are the facts that when they experience symptom relief, many patients tend to drop out of therapy or discontinue medication. As I have argued, depression is a behavioral disorder and treatments often relieve the symptom of depression, the affective component, without modifying the underlying behaviors. This is why the most positive outcomes are observed using a combination of medication and therapy. Medication alleviates the symptoms of the disorder while the patient works to modify their behavior through therapy.

Medication alone is not a cure for depression, but even the most effective medical treatments have low efficacy. For a discussion of the complications associated with gaging the effectiveness of antidepressants, please check out this article from the National Institute of Mental Health at http://www.nimh.nih.gov/about/director/2011/antidepressants-a-complicated-picture.shtml.

19 Martin, R. A. (2010). The psychology of humor: An integrative approach. Academic press.

20 McGraw, A. P. and Warren, C. (2010). Benign violations making immoral behavior funny. *Psychological Science*, 21(8), 1141–9.

21 On the BBC's *The Graham Norton Show*. Although I understand the interview may have been recorded in 2014. Clip available at: www.bbc.co.uk/programmes/p02s0bpr.

22 von Radowitz, John, "Revealed: The funniest Joke in the world," *The Guardian*, October 3, 2002, available at http://www.theguardian.com/uk/2002/oct/03/3. A link to this study may be found by following a link on the LaughLab website at http://laughlab.co.uk/ or, directly, at https://richardwiseman.files.wordpress.com/2011/09/ll-final-report.pdf.

23 Wiseman, LAUGHLAB: THE SCIENTIFIC SEARCH FOR THE WORLD'S FUNNIEST JOKE, 2002, p. 3

24 Paul Kelbie, "How Spike Milligan had the last laugh - and the best," *The Independent*, June 9, 2006, available at http://www.independent.co.uk/news/uk/this-britain/how-spike-milligan-had-the-last-laugh—and-the-best-481640.html.

25 Wiseman, LAUGHLAB, p. 14

26   Randy Lewis, "George Harrison Memorial Tree killed ... by bee-
      tles; replanting due," *Los Angeles Times*, July 21, 2015, a http://www.
      latimes.com/entertainment/music/posts/la-et-ms-george-harrison-
      tree-beetles-replant-20140721-story.html.
27   May 22, 2005, available at http://www.nytimes.com/2005/05/22/fash-
      ion/sundaystyles/seriously-the-joke-is-dead.html.
28   Kifner, John, "No Joke! 37 Years After Death Lenny Bruce Receives
      Pardon," *New York Times*, December 24, 2003, http://www.nytimes.
      com/2003/12/24/nyregion/no-joke-37-years-after-death-lenny-
      bruce-receives-pardon.html.
29   See, for example:
      Kahneman, D. (2011). *Thinking, Fast and Slow*. Macmillan.
30   For example:
      Morin, A. and Hamper, B. (2012). "Self-reflection and the inner
      voice: activation of the left inferior frontal gyrus during perceptual
      and conceptual self-referential thinking." *The Open Neuroimaging
      Journal*, 6, 78.
31   Chan, Y. C., Chou, T. L., Chen, H. C., and Liang, K. C. (2012).
      "Segregating the comprehension and elaboration processing of ver-
      bal jokes: an fMRI study." *NeuroImage*, *61*(4), 899–906.
      Chan, Y. C., Chou, T. L., Chen, H. C., Yeh, Y. C., Lavallee, J. P., Liang,
      K. C., and Chang, K. E. (2013). "Towards a neural circuit model of
      verbal humor processing: An fMRI study of the neural substrates of
      incongruity detection and resolution." *Neuroimage*, 66, 169–76.
      Marinkovic, K., Baldwin, S., Courtney, M. G., Witzel, T., Dale, A.
      M., and Halgren, E. (2011). "Right hemisphere has the last laugh:
      neural dynamics of joke appreciation." *Cognitive, Affective, and
      Behavioral Neuroscience*, *11*(1), 113–130.
      Samson, A. C., Zysset, S., and Huber, O. (2008). "Cognitive humor
      processing: different logical mechanisms in nonverbal cartoons—an
      fMRI study." *Social Neuroscience*, *3*(2), 125–40.
      Chan, Y. C., Chou, T. L., Chen, H. C., Yeh, Y. C., Lavallee, J. P., Liang,
      K. C., and Chang, K. E. (2013). "Towards a neural circuit model of
      verbal humor processing: An fMRI study of the neural substrates of
      incongruity detection and resolution." *Neuroimage*, 66, 169–76.
32   Like the prefontal cortex, the corpus callosum is also not fully
      developed until one's mid-twenties—which is one reason why a lot
      of comedians don't like performing at all-age shows in favor of the
      over twenty-one crowd.
33   Brown, W. S., Paul, L. K., Symington, M., and Dietrich, R. (2005).
      "Comprehension of humor in primary agenesis of the corpus cal-
      losum." *Neuropsychologia*, *43*(6), 906–16.

34   Not to say the guy's failure to get my sarcasm means he was demented, but see:
     Kipps, C. M., Nestor, P. J., Acosta-Cabronero, J., Arnold, R., and Hodges, J. R. (2009). "Understanding social dysfunction in the behavioural variant of frontotemporal dementia: the role of emotion and sarcasm processing." *Brain*, *132*(3), 592–603.

35   These include Asperger's syndrome, language center impairment, schizophrenia, and right hemisphere damage.

36   Brown, W. S. Paul, L. K., Symington, M., and Dietrich, R. (2005). "Comprehension of humor in primary agenesis of the corpus callosum." *Neuropsychologia*, *43*(6), 906–16.
     Coulson, S. and Williams, R. F. (2005). Hemispheric asymmetries and joke comprehension. *Neuropsychologia*, *43*(1), 128–141.
     Gardner, H., Ling, P. K., Flamm, L., and Silverman, J. E. N. (1975). "Comprehension and appreciation of humorous material following brain damage." *Brain*, *98*(3), 399–412.
     Shammi, P., and Stuss, D. T. (1999). "Humour appreciation: a role of the right frontal lobe." *Brain*, *122*(4), 657–66.

37   Braun, C. M., Lussier, F., Baribeau, J. M., and Ethier, M. (1989). "Does severe traumatic closed head injury impair sense of humour?" *Brain Injury*, *3*(4), 345–54.

38   Samson, A. C., and Hegenloh, M. (2010). "Stimulus characteristics affect humor processing in individuals with Asperger syndrome." *Journal of Autism and Developmental Disorders*, 40(4), 438–47.

39   Darwin, C., Ekman, P., and Prodger, P. (1998). "The expression of the emotions in man and animals." Oxford University Press. and Provine, R. R. (2001). *Laughter: A Scientific Investigation*. Penguin.

40   Provine, R. R. (2001). *Laughter: A Scientific Investigation*. Penguin.

41   "Research done by Patricia Simonet at Sierra Nevada College in Lake Tahoe looked at laughter sounds in dogs. Simonet's team investigated the question by standing in parks with a parabolic microphone that allowed them to record the sounds that dogs made while playing from a distance. In describing the laughter sounds of dogs she says that, 'To an untrained human ear, it sounds much like a pant, 'hhuh, hhuh.' When the recordings were analyzed she found that that this exhalation bursts into a broader range of frequencies than does regular dog panting. She confirmed the positive effects of this laugh sound in an experiment on fifteen puppies, which romped for joy simply upon hearing the recorded canine laugh. More recently she was able to show that these same sounds helped to calm dogs in an animal shelter." Stanley Coren, "Do Dogs Laugh? Animals make laugh-like sounds when they are tickled or playing,"

Psychology Today, November 22, 2009, https://www.psychologyto-day.com/blog/canine-corner/200911/do-dogs-laugh.

42   James Gorman, "Dogs May Laugh, but Only Cats Get the Joke," *The New York Times*, September 5, 2006, available at http://www.nytimes.com/2006/09/05/science/05side.html.

43   Panksepp, J. (2007). "Neuroevolutionary sources of laughter and social joy: Modeling primal human laughter in laboratory rats." *Behavioural Brain Research,182*(2), 231–44.

44   Robert Provine has shown that, in conversation, non-humorous laughter punctuates verbal speech in a similar way.

45   "Popular aphorisms such as 'It takes forty-three muscles to frown and only seventeen to smile" claiming that a greater number of facial muscles are needed to produce a frown than to generate a smile are one such snippet of homespun wisdom; they have been a part of our cultural landscape for so long that no one now knows where it all began . . . . While there are no scientific basis for such claims (technically you need twelve muscles to do a true smile and eleven to frown . . . but even those numbers are arguable), such sayings serve to convey a very important message: smiling makes us feel happier." Excerpted from "Does it really take more muscles to frown than to smile?" Laughter Online University. http://www.laughter-onlineuniversity.com/true-false-takes-43-muscles-frown-17-smile/.

46   Martin, R. A. (2010). "The psychology of humor: An integrative approach." Academic press.

47   Importantly, recovering from such gasps means longer exhalations, ridding the lungs of residual air, and enriching the blood with replenished supplies of oxygen compared to pre-laughter.

48   The parasympathetic branch of the autonomic nervous system controls the lacrimal glands via the neurotransmitter acetylcholine through both the nicotinic and muscarinic receptors. When these receptors are activated, the lacrimal gland is stimulated to produce tears.

49   Interestingly, there is very little research out there on the function of applause. What I offer here is my interpretation, and one possibility (we applaud to make use of an automatic muscle contraction). However my friend William, who has been helping me, insists that I include his idea as well. To him, applause is a signal to other members of our group that we are in a safe place, as the noise would otherwise attract predators or enemies, or an invitation to play. I would suggest that it probably has multiple functions and don't entirely disagree with him. However, I do disagree with what he is wearing.

50   Provine, R. R. (1996). Laughter. *American Scientist*, 38–45.

51   See, e.g., McGraw, P. and Warner, J. (2014). *The Humor Code: A Global Search for What Makes Things Funny*. Simon and Schuster.

52   Yale Encyclopedia of Alternative Medicine, by Ken R. Wells, available at http://findarticles.com/p/articles/mi_g2603/is_0004/ai_2603000435.

53   Angelman Syndrome

54   The podcast can be found at http://www.radiolab.org/story/91588-laughter/.

55   Six thousand clubs in sixty-five countries according to www.laughteryoga.org/english/training/details/129, or—more recently—eight thousand clubs in one hundred countries according to http://www.laughteryoga.org/english.

56   Stutzer, A. and Frey, B. S. (2006)." Does marriage make people happy, or do happy people get married?" *The Journal of Socio-Economics*, *35*(2), 326–47.

57   Article in the Aspen Idea Blog, June 28, 2015. The Aspen Institute. http://www.aspeninstitute.org/about/blog/us-surgeon-general-vivek-murthy-gun-violence-happiness-more.

58   Abel, E. L. and Kruger, M. L. (2010). "Smile intensity in photographs predicts longevity." *Psychological Science*, 21(4), 542–44.

59   Kringelbach, M. L. and Berridge, K. C. (2009). "Towards a functional neuroanatomy of pleasure and happiness." *Trends in cognitive sciences*, 13(11), 479–87.

60   Olive, M. F., Koenig, H. N., Nannini, M. A., and Hodge, C. W. (2001). "Stimulation of endorphin neurotransmission in the nucleus accumbens by ethanol, cocaine, and amphetamine." *J Neurosci,* 21(23), RC184.

61   These data are aggregated across several studies, one of which was published as:
     King, B. and Alexander, G. (2000). "Pain sensitivity and individual differences in self-reported sexual behavior. *Journal of Comparative Psychology*, 193–9.

62   Dunbar, R. I., Baron, R., Frangou, A., Pearce, E., van Leeuwin, E. J., Stow, J., . . . and Van Vugt, M. (2011). "Social laughter is correlated with an elevated pain threshold." Proceedings of the Royal Society of London B: Biological Sciences, rspb20111373.

63   One example: Trice, A. D. and Price-Greathouse, J. (1986). "Joking under the drill: A validity study of the coping humor scale." *Journal of Social Behavior and Personality*.

64   Morse, D. R. (2006). Use of humor to reduce stress and pain and enhance healing in the dental setting. *Journal of the New Jersey Dental Association,* 78(4), 32–36.

65   Stuber, M., Hilber, S. D., Mintzer, L. L., Castaneda, M., Glover, D., and Zeltzer, L. (2009). Laughter, humor and pain perception in

children: a pilot study. *Evidence-based complementary and alternative medicine*, *6*(2), 271–6.

66   Christopher Bergland, "Cortisol: Why 'The Stress Hormone' Is Public Enemy No. 1," Psychology Today, January 22, 2013, available at https://www.psychologytoday.com/blog/the-athletes-way/201301/cortisol-why-the-stress-hormone-is-public-enemy-no-1.

67   Fry, W. F., and Savin, W. M. (1988). Mirthful laughter and blood pressure.

68   Miller, M., Mangano, C., Park, Y., Goel, R., Plotnick, G. D., and Vogel, R. A. (2006). Impact of cinematic viewing on endothelial function. *Heart*, *92*(2), 261–2.

69   Berk, L. S., Tan, S. A., Nehlsencannarella, S., Napier, B. J., Lewis, J. E., Lee, J. W., . . . and Fry, W. F. (April 1988). "Humor associated laughter decreases cortisol and increases spontaneous lymphocyte blastogenesis." *Clinical Research*, vol. 36, no. 3, pp. A435-A435. Slack Inc.

70   Bennett, M. P., and Lengacher, C. (2006). "Humor and laughter may influence health: II. Complementary therapies and humor in a clinical population." *Evidence-Based Complementary and Alternative Medicine*, *3*(2), 187–90.
      Berk, L. S., Tan, S. A., and Berk, D. (2008). Cortisol and Catecholamine stress hormone decrease is associated with the behavior of perceptual anticipation of mirthful laughter. *The FASEB Journal*, *22*(1_ MeetingAbstracts), 946–11.

71   Berk, L. S., Tan, S. A., Napier, B. J., and Eby, W. C. (1989, January). Eustress of Mirthful Laughter Modifies Natural-Killer Cell-Activity, *Clinical Research* (Vol. 37, No. 1, pp. A115-A115). Slack Inc.

72   Field, T., Hernandez-Reif, M., Diego, M., Schanberg, S., and Kuhn, C. (2005). "Cortisol decreases and serotonin and dopamine increase following massage therapy." *International Journal of Neuroscience*, *115*(10), 1397–1413.

73   Thoma, M. V., La Marca, R., Brönnimann, R., Finkel, L., Ehlert, U., and Nater, U. M. (2013). T"he effect of music on the human stress response." *PloS one*, 8(8), e70156.

74   Lane Anderson, "Mind Your Body: Dance Yourself Happy," *Psychology Today*, July 1, 2011, available at https://www.psychology-today.com/articles/201008/mind-your-body-dance-yourself-happy.

75   Patrick, C. J., Fowles, D. C., and Krueger, R. F. (2009). "Triarchic conceptualization of psychopathy: Developmental origins of disinhibition, boldness, and meanness." *Development and Psychopathology*, *21*(03), 913–38.

76 See, for example, Martin E. P. Seligman's "Learned Helplessness" for a review of this research by himself and others. *Annual Review of Medicine*, Vol. 23: 407–12 (February 1972).

77 Arnsten, A. F. (2009). "Stress signaling pathways that impair prefrontal cortex structure and function." *Nature Reviews Neuroscience*, *10*(6), 410–22.

78 Skinner, E. and Greene, T. (2008). "Perceived control: Engagement, coping, and development." *21st Century Education: A Reference Handbook*, *1*, 121–30.

79 Bennett, M. P., Zeller, J. M., Rosenberg, L., and McCann, J. (2003). "The effect of mirthful laughter on stress and natural killer cell activity." *Nursing Faculty Publications*, 9.

80 List is available here: http://www.gq-magazine.co.uk/entertainment/articles/2011–03/09/-gq-jokes-best-celebrity-laughs-all-time.

81 Azim, E., Mobbs, D., Jo, B., Menon, V., and Reiss, A. L. (2005). "Sex differences in brain activation elicited by humor." *Proceedings of the National Academy of Sciences of the United States of America*, 102(45), 16496–501.

82 King, B. E. and Alexander, G. M. (2000). "Pain sensitivity and individual differences in self-reported sexual behavior." *Journal of Comparative Psychology*, 114(2), 193.

83 Winters, J., Christoff, K., and Gorzalka, B. B. (2009). "Conscious regulation of sexual arousal in men." *Journal of Sex Research*, 46(4), 330–343.

84 Whipple, B. and Komisaruk, B. R. (1985). "Elevation of pain threshold by vaginal stimulation in women." *Pain*, 21(4), 357–67.
Whipple, B. and Komisaruk, B. R. (1988). "Analgesia produced in women by genital self-stimulation." *Journal of Sex Research*, 24(1), 130–40.

85 Linehan, M. M. (2013). Dialectical behavior therapy. Milton H. Erickson Foundation.

86 Ando, V., Claridge, G., and Clark, K. (2014). "Psychotic traits in comedians." *The British Journal of Psychiatry*, 341–5.

87 Moore, A. and Gibbons, D. (1987). *Watchmen* (Book club ed.). New York: DC Comics.

88 Anxiety disorders are the most common mental illness in the US, affecting forty million adults in the United States age eighteen and older, or 18 percent of the population. (Source: National Institute of Mental Health, http://www.nimh.nih.gov/health/statistics/prevalence/any-anxiety-disorder-among-adults.shtml.

89 World Health Organization (WHO) report quoted in BBC Online January 9, 2001; see also "Global depression statistics" July 26,

2011, BioMed Central, available at http://www.sciencedaily.com/releases/2011/07/110725202240.htm.

90 Peter Gray, "The Decline of Play and Rise in Children's Mental Disorders," *Psychology Today*, January 26, 2010, https://www.psychologytoday.com/blog/freedom-learn/201001/the-decline-play-and-rise-in-childrens-mental-disorders.

91 This is clearly a joke reference to the plot of the 2001 Disney/Pixar movie *Monsters Inc.* Please don't sue me.

92 As recounted in *Mystery of Happiness* [Motion picture on DVD]. (1996). ABC News.

93 Carlson, R. (1997). "Don't sweat the small stuff—and it's all small stuff: Simple ways to keep the little things from taking over your life." New York: Hyperion.

94 David Granirer's work can be found at www.standupformental-health.com.

95 You can find information on Daniel and his work at www.improveyoursocialskills.com.

96 The AATH provides information about humor and laughter, promotes research, and supports programs that utilize therapeutic humor. Check them out at www.aath.org.

97 Not that we'd need children to enjoy a day at Disney. But I must say that there are some parks that I don't know if I'd be comfortable visiting without kids in tow. I'd love to go to a LEGOLAND park, but that seems like the kind of place I'd have to borrow some kids to take.

98 Appropriately hosted at www.patchadams.org.

99 Ventis, W. L., Higbee, G., and Murdock, S. (2001). "Using Humor in Systematic Desensitization to Reduce Fear." *The Journal of General Psychology*, 241–53.

100 I am teasing Ohio here; I am very grateful for the opportunity the psychology department at Bowling Green State University gave me to complete my studies.

101 Shahidi, M., Mojtahed, A., Modabbernia, A., Mojtahed, M., Shafiabady, A., Delavar, A., and Honari, H. (2011). "Laughter yoga versus group exercise program in elderly depressed women: a randomized controlled trial." *International Journal of Geriatric Psychiatry*, 26(3), 322–7.

102 Dolgoff-Kaspar, Rima, et al. "Effect of laughter yoga on mood and heart rate variability in patients awaiting organ transplantation: A pilot study." *Alternative Therapies in Health and Medicine*, 18.5 (2012): 61–66.

103 Yazdani, M., Esmaeilzadeh, M., Pahlavanzadeh, and S., and Khaledi, F. (2014). "The effect of laughter yoga on general health

among nursing students." *Iranian Journal of Nursing and Midwifery Research*, 19(1), 36.

104 I first learned of the phrase "incest is best" on an episode of *The Jerry Springer Show*. Now you've learned it too. You can't unlearn it.

105 If you do catch me on the road, be sure and say hi!

106 I figured the end of the book should have its own endnote. Is anyone reading these anyway?

ACKNOWLEDGMENTS

In preparing this work, I owe a great deal of thanks to my friend William von Sydow, often referred to as my lawyer, for his valuable assistance. I would also like to thank comedian Jason Goodall, also referred to as my financial adviser, for his commentary on the first draft of the manuscript. Speaking of commentary, I am grateful to Dr. Peter McGraw and, while I may not fully understand it, his inspiring theory, and for his consultation and feedback early on.

I would like to thank all the comedians and comedy fans who answered my calls, allowed me to interview them, or provided me with a story to use. I list them in Appendix II. Special thanks goes to comedians Greg Dean, Ariel Elias, and Katrina Brown for their commentary on various sections of the draft.

I would like to thank every booker to ever give me stage time, as well as every show producer and every host I've worked with. Likewise, I would like to thank every person who has ever attended one of my shows, seminars, or talks.

Finally, I would like to thank everyone who has ever made me laugh. And to those who have laughed with me, you're welcome.